S0-DUU-822

REASSESSMENT OF MUSCLE HOMOLOGIES AND NOMENCLATURE

[**in Conservative Amniotes,**
The Echidna, *Tachyglossus*,
The Opossum, *Didelphis*,
and The Tuatara, *Sphenodon*]

First Edition

Adelaide (Heidi) Frost Ellsworth, Ph.D.
University of Connecticut

ROBERT E. KRIEGER PUBLISHING COMPANY HUNTINGTON, NEW YORK

ORIGINAL EDITION 1974

Printed and Published by
ROBERT E. KRIEGER PUBLISHING CO., INC.
BOX 542, HUNTINGTON, NEW YORK 11973

Library of Congress Card Number 000000
ISBN 0-88275-158-1

Printed in the United States of America

Library of Congress Cataloging in Publication Data

Ellsworth, Adelaide Heide Frost.
Reassessment of muscle homologies and nomenclature in conservative amniotes.

Bibliography: p.
1. Muscles. 2. Tachyglossus. 3. Didelphis. 4. Tuatara. I. Title.
QL831.E44 591.1'852 74-8907
ISBN 0-88275-158-1

REASSESSMENT OF MUSCLE HOMOLOGIES AND NOMENCLATURE IN CONSERVATIVE AMNIOTES, THE ECHIDNA, *Tachyglossus*, THE OPOSSUM, *Didelphis*, AND THE TUATARA, *Sphenodon*, PART I: PELVIC MUSCULATURE [with] DRAWINGS, PLATES AND APPENDIX

Adelaide Frost ELLSWORTH, Ph.D.
New York University, 1972

The exploratory dissection of superficial and medium depth muscles on the pelvic region of echidnas, opossums, and tuataras, reaffirms the few differences and the many similarities in structure and form in the moving apparatus of primitive mammals and their connection to reptiles. The detailed descriptions of the muscular systems testify to the fact that comparative patterns are a repeated set of relationships. The shift of the ilium to the horizontal position in *Didelphis* and *Tachyglossus* is commonly known to explain the paths of incidental muscle migration and the occasional accompanying displacement of nerves. The basic muscle pattern has been maintained in the rhynchocephalian, the monotreme, and the marsupial.

Specifically, because of its relation to other muscles in topography, the pubi-tibialis of reptiles (Gadow, 1882, and others) which is supposed to have no complement in mammals, has been called in this investigation a Musculus adductor; its nerve supply comes from the Plexus lumbalis in tuatara.

The early Triassic fossils of the monotremes, marsupials, and the original rhynchocephalians show fundamental resemblances which have not been blotted out. There can exist homology because there is direct evidence in the dissections presented, which shows anatomic conditions in degrees of relationships, and genetic parallels among conservative organisms. This preservativeness in the opossum, *Didelphis*, is generally known to be due to its ability to withstand changes in climate, diet, and environment, to its multiplicity of young and a short life span; the genera is semi-ubiquitous throughout North America. The maintenance of form in the echidna, *Tachyglossus*, is believed to be due to its isolation on the continent of Australia and associated islands. This primitive animal is thriving right now in this area (Simpson, 1969. Personal communication). The tuatara, *Sphenodon*, with the long life span, according to the same source, seems due in the future to extinction. Slowly they are being removed from the islands around New Zealand; what remain have held to the muscle models (seen from fossil bones) of their ancestors. The fossil patterns appear universal in all these conservative animals.

This new view of muscle correspondence adds somewhat to the simplification of the nomenclature. Most muscles, bones and nerves researched are given the designations put forth in the 1968 edition of the Nomina Anatomica Veterinaria.

ACKNOWLEDGMENTS

With the kind permission of the Australian Government the protected monotreme echidnas were sent preserved to the United States. *Tachyglossus setosus* was shipped from the Hobart area, Tasmania. Two others, *Tachyglossus aculeatus*, were obtained from Clayton, Victoria, through the courtesy of Dr. E. H. M. Ealey. Embalmed species of *Didelphis marsupialis*, the American opossums, whose subspecies were unknown (location of capture could be roughly approximated as from Florida and North Carolina) were shipped from Carolina Biological Supply Company.

With Governmental authorization, New Zealand transported two living *Sphenodon punctatus*, tuataras, to us for this muscle research. The disarticulated skeleton of the echidna, *Tachyglossus*, and tuatara, *Sphenodon*, were loaned by the American Museum of Natural History; that of the opossum, *Didelphis*, and the mounted specimen were purchased from Carolina Biological Supply Company. The Biology Department of New York University permitted use of their intact echidna skeleton, as well as crocodile and lizard whole animal mounts. An articulated tuatara was borrowed from the Museum of Comparative Zoology at Cambridge.

I wish to thank Dr. Ralph Wetzel of the University of Connecticut, who helped obtain all animals; this Biological Sciences Department gave graduate space for the study, as well as use of bones, skeletons and chemicals. One specimen of tuatara was kept alive there for six weeks to study reptilian locomotion; it was finally sacrificed and some organs used for physiological study. A live echidna has survived importation thirty-six months.

I wish also to acknowledge the advice of Dr. Eva Weinreb of New York University for some of the early preparation of this manuscript. Dr. David Klingener of the University of Massachusetts reviewed the opossum muscles; Dr. Farish Jenkins of the Peabody Museum at Yale University checked the echidna dissections. This investigation was approved by Dr. Alfred Romer of the Museum of Comparative Zoology at Harvard; it was initiated by Dr. Ralph Wetzel, Mammalogist, of the University of Connecticut. Subsequently, Dr. Arthur M. Crosman of the Graduate Biology Department of New York University supervised in the final study of primitive amniotes. I deeply appreciate his patient encouragement and helpful advice. Dr. Thomas Nicholson, of the American Museum of Natural History, authorized quotations from several of the Museum's Bulletin publications. I am grateful to Miss Brenda Dion and Mrs. Judith Peck for their help in typing the manuscript; also Mrs. Winifred Hundgen and Miss Susanne Kremer, who assisted with the drawings.

It is with gratitude I dedicate this manuscript to my patient family, without whose support the detailed study could not have been undertaken—and to Dr. Gladys Martyko for her continued encouragement; to Dr. M. J. Kopac for his interest in the supervision of the research.

TABLE OF CONTENTS

INDEX TO FIGURES

INDEX TO PLATES

INDEX to SPECIFIC MUSCLES of THE PELVIS

I INTRODUCTION

Mammalian anatomy shows many reptilian characteristics. As Dr. Romer has put it:

> The sprawled quadrupedal pose, with a broad trackway, is common to nearly all the varied cotylosaurs . . . and is persistently present in the early mammal-like forms of the order Pelycosauria, which were the dominant animals of the early Permian . . . Of the persistently quadrupedal reptiles it is the mammal-like Therapsida which most greatly modified and improved their locomotor apparatus. In these forms there developed a highly efficient, four-footed, running gait which in advanced members was not far from that of typical mammals . . . the elbow turned back, the knee forward, with numerous structural changes in girdle and limb bone construction, to give a narrow trackway, with the limbs providing direct vertical support to the body. The pareiasaur trend in this direction was related to problems of weight support . . . the selective factor in the development of this posture in mammal-like forms appears to have been that of increased speed. A longer stride is possible with the limbs in this fore-and-aft pose, and muscular energy can be more efficiently used directly for propulsion instead of being wasted in the mere maintenance of posture necessary in the sprawled position.[1]

It is the purpose of this investigation to describe the comparative pelvic musculature anatomy of conservative living mammaliae, *Tachyglossus* and *Didelphis marsupialis*, and attempt to relate this musculature to that of the only living reptilian species of Rhynchocephalia, *Sphenodon* (Fig. 2). Three primitive amniotes were studied in detail, the echidna (*Tachyglossus*), the opossum (*Didelphis*), and the tuatara (*Sphenodon*). In a study of musculature in such research, skeletology must be considered in terms of origins and insertions on fixed and moving parts.

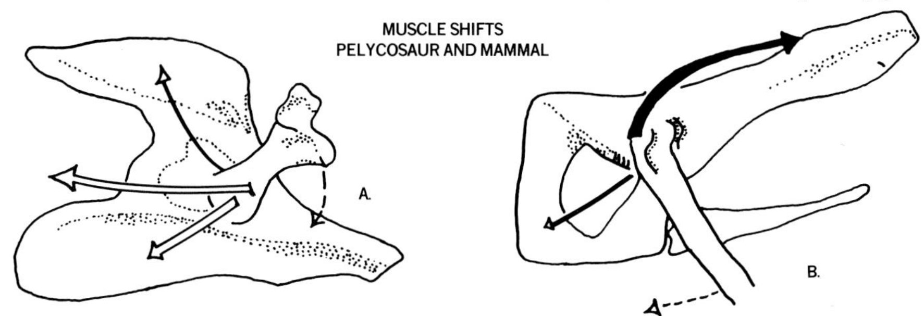

Fig. 1. Diagram to show the differences in the musculature of the hip region between a pelycosaur (a) and a mammal (b). In A the principal muscles pulling the femur downward and backward (shown in heavy lines) attach to the tail and ventral part of the pelvic girdle. In B these muscles are reduced and replaced to a considerable extent by muscles attached to the greater trochanter and inserting on the ilium.

Redrawn from A.S. Romer: Vertebrate Paleontology 1966. Courtesy of University of Chicago Press. Reprinted through permission of the publisher, Fig. 302,

I have come to believe that the posterior muscles of these three genera are very near the

[1]Romer, A. 1968a. The Osteology of the Reptiles. pp. 337-339, The University of Chicago Press, Chicago, Courtesy of the University of Chicago Press.

original pattern of the common ancestral type, reconstructed from the fossil record (Romer, 1922; 1923). The muscle innervations are nearly identical except occasionally where there is loss or addition of certain twigs; the origins and insertions show but slight changes in muscle fascicle relationships. The pelvic, caudal, and posterior appendages have slightly different shpes, which have obviously developed as a result of progressive activity and motion. Bone was laid down anew a little at a time, as muscle tension and use reconverted part of each pelvis, shank and leg, and nerves followed and formed new neuromuscular junctions. But despite minor changes in figure, from what is known concerning evolution, each conservative animal has maintained itself as a whole organism, each in its best possible form in deference to its habitat and environment. There is the same basic muscle pattern in all the genera under study. Therefore, it is logical to assign the same nomenclature to similar muscles of all three animals, noting, however, that in worth, force, import, number, amount, value and use there is not exact muscle equivalence.

In this comparative study on myology, emphasis is placed on the observed evolutionary arrangement of parts, in order to explain progression in form and structure of muscles in primitive mammals such as the prototherian monotremes and metatherian marsupials. In the egg-laying order rhynchocephalia under scrutiny, the significant factors are the striking myological segmentation, the abundance of the connective tissue binding the muscle attachments, the distinct dorsal and separate ventral muscle masses and slips, and generally, their basic somite innervations. These illustrate the primitive muscles found in all three conservative amniotes.

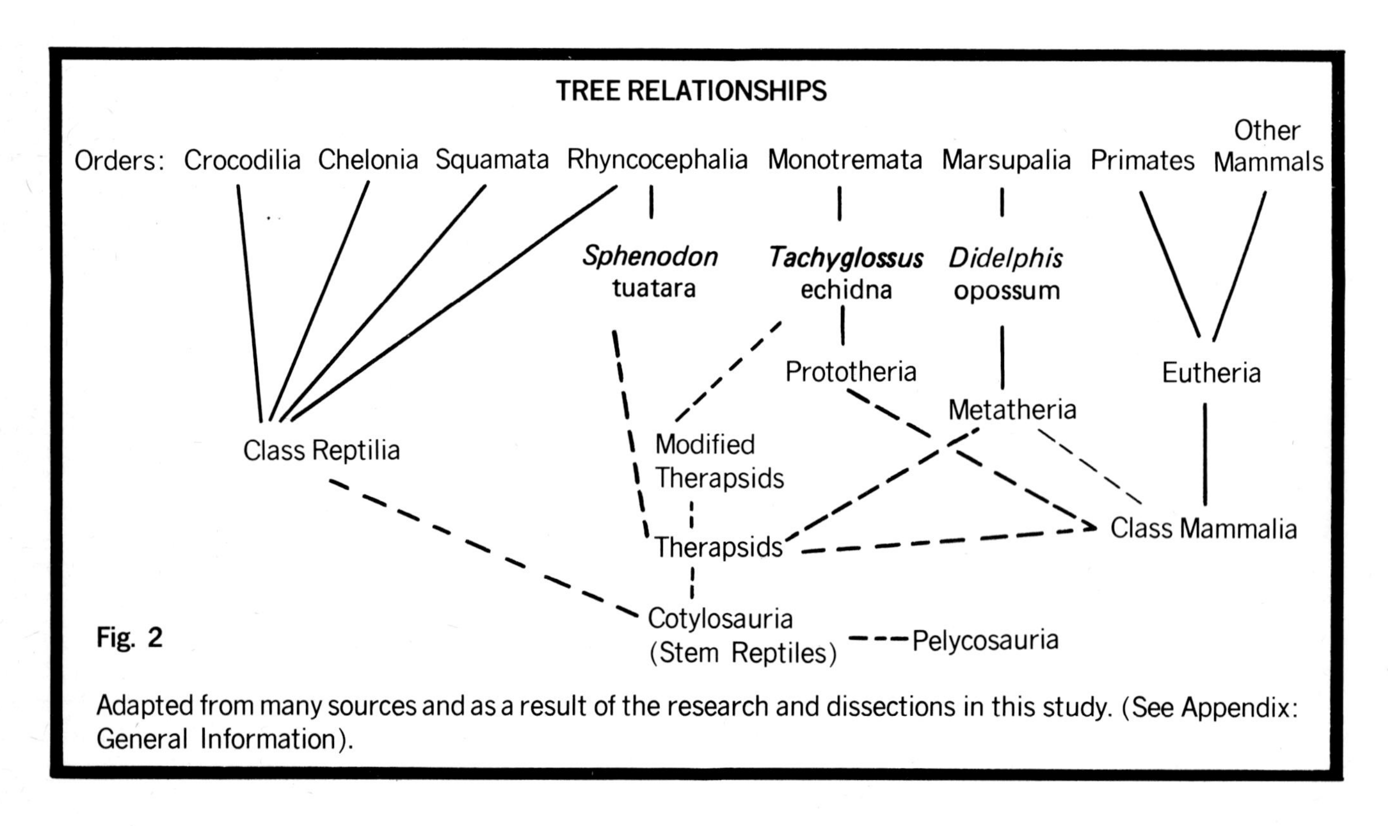

Fig. 2

Adapted from many sources and as a result of the research and dissections in this study. (See Appendix: General Information).

II MATERIALS AND METHODS

Mature and some immature species of echidna were dissected; one example of *T. setosus*, and two of *T. aculeatus* of approximately the same age were used. Three preserved specimens of the opossum, *D. marsupialis*, also were studied; two *S. punctatus* were carefully researched and examined. Both males and females were among the animals; ages were uncertain. Intact and disarticulated skeletons were employed for accompanying bone investigations.

A complete unilateral dissection of the pelvic musculature was performed on each animal. Systematic study involved the relationship of each muscle to underlying bone with reference to origin, insertion and position. Nerves were followed as far as possible to each muscle fascicle. Every dissection was photographed in color at different levels of sheet muscle. The purpose of this was to insure reference for further study, and to make laboratory prints that looked real. Laboratory specimens were kept between workouts in embalming fluid, and each was wrapped in cheesecloth soaked in the liquid.

Detailed life-sized drawings of the pelvic regions of the echidna, the opossum, and tuatara are included for correlation. Color photographs of muscle dissections were made as follows: a 35 mm Mamiya Auto-Lux camera (Japanese) was employed, using Kodak Kodacolor-X, CX135-20 film; the pictures were taken with flood lights generally through magnifying lenses, and the negatives were enlarged to 5X7.

The nomenclature throughout this report is generally consistent with NOMINA ANATOMICA (third edition, 1968, EXCERPTA MEDICA FOUNDATIO, Amsterdam); also carefully followed was NOMINA ANATOMICA VETERINARIA (first edition, 1968, WORLD ASSOCIATION OF VETERINARY ANATOMISTS, Vienna).

The nature of this study necessitates that the discussion for the same muscle in the echidna and the opossum follow the observation, relating each bundle to that of tuatara. The findings about other reptiles living today, whose myology has been extensively investigated, are reported in sequential notes. This seems in order, to better correlate information in every case. If the muscles collectively were brought into the discussion, each would be taken out of context and the results would prove confusing and cumbersome to the reader.

The description and comparison of the superficial pelvic girdle and thigh muscles have been divided into the following classifications:

The Extensor System

- Iliacus Group
- Gluteal Group
- The Sartorius Muscle
- Quadriceps Femoris Group

The Flexor System

- Adductor Group
- Hamstring Group

III OBSERVATIONS

The muscles of the pelvic girdle and thigh were dissected in order to determine what muscles are basically the same in reptiles, monotremes and marsupials. In the tibial region only a few are mentioned, and these are important because they relate to the changed posture of mammals from the therapsid to the marsupial pose. The altered angle of the girdle and the long bones of the appendicular skeleton of the hind limb reveal how conservative a primitive animal can be. In addition, it was found that the American opossum, *Didelphis*, a marsupial, bears out its label as a generalized organism. It is not necessarily true that such an organism is primitive, but it has been shown from the fossil records that this is a valid conclusion in *Didelphis*. *Tachyglossus*, the echidna, a monotreme, is also a primitive arrival. Simpson (1964) considers it a highly modified surviving therapsid reptile, only a mammal because the echidna fits the definition of one.[2]

It has been shown that *Sphenodon*, tuatara, the only surviving rhynchocephalian, is very close to mammal forms in its myology. The flexors are of enormous bulk and well developed; they abound in accessory tendons but remain in the generalized condition in the hind limb region.

The observations are confined to the superficial extensor and flexor muscles. The extensor system is divided into the iliacus, gluteal, the sartorius muscle, and quadriceps femoris groups.

The iliacus unit includes the Musculi iliopsoas (part of the multiple psoas) and the pectineus. These muscles are innervated by branches of the Plexus lumbosacralis. The gluteal assemblage is made up of the Musculi glutei and the femorococcygeus. These muscles collectively are innervated by parts of the Nervi femoris and cruris. The glutei vary in the number of their parts and in some animals can be classified in layers, but in the primitive mammals what is seen are superficial and various deep arrangements which differ in position in the subclasses. The nerves follow the longitudinally divided tissue. The Musculus sartorius, a single unit, serving to extend or straighten the leg, has Plexus lumbalis nerve twigs attached. Other superficial extensors, the quadriceps femoris group, contain the Musculi rectus femoris and the vasti. They are supplied with nerves from the Nervus femoralis and its branches.

The flexor system is comprised of the adductor group and the recognizable unit known as the hamstrings. The former include the Musculi gracilis and the true adductores. Nervus obturatorius supplies the innervations. The muscles in the hamstring group are composed of the Musculi biceps femoris, semimembranosus, semitendinosus; (piriformis = femorococcygeus = femorocaudalis ?). They receive divisions from the Nervus tibialis and the more terminal branch of the sciatic nerve, the Nervus ischiadicus in *Didelphis*. In *Tachyglossus* and *Sphenodon*, the postsacral nerves are involved, mostly; *i.e.*, Nervi caudalis (Nerve Patterns, pp. 17 ff.).

[2] Simpson, G. G. 1964. The Meaning of Evolution. Yale University Press, New Haven, Conn., p.66.

A. THE EXTENSOR SYSTEM

1. Iliacus Group

These muscles are innervated by branches of the Plexus lumbalis.

MUSCULUS ILIOPSOAS

Didelphis (Plate 1-w: Fig. 3)

Origin: The iliacus portion from the apex and iliac fossa of the ilium; the fan-like psoas minor from the lumbar vertebrae only, on a common tendon with the psoas major from the lumbar and sacral vertebrae. The latter overlays the head of the femur.

Insertion: By blending of the muscle fibers into a tendon that wraps around and goes into the trochanter minor.

Tachyglossus (Plate 2-c: Fig. 4)

Origin: The iliacus part from the ventral lateral surface of the ilium; the long psoas major-minor segment blending from three lumbar and sacral vertebrae.

Insertion: Together, the iliacus constituent on the border of the femur; the psoas major-minor by triangular points of attachment on a tendon, and some fibers into the dorsal neck of the femur between the trochanters.

Sphenodon (Plates 6-p;16-z; 28a: Figs. 10-p; 11)

Origin: A psoas component comes backward from within; the iliacus portion extends entirely from the inner surfaces of the dorsal side of the ventral pelvic plate, the dorsal side of the epipubis, the pubis, the ischium and the Membrana obturatoria (Plate 27). The Musculus pectineus is an element of this complex in tuatara, *Sphenodon*.

Insertion: Fibers, sometimes, are attached to the vertebrae in front of the sacrum (Musculus psoas?). The iliac piece comes forward to the medial side of the proximal part of the femur. All wrap and make a circle around the anterior pubic-iliac angle.

It is in the opossum, *Didelphis*, and echidna, *Tachyglossus*, that the psoas minor is the larger of the two psoas muscles. In both animals these muscles are elongated and curved. *Sphenodon's* (tuatara's) entire sheet with the Musculus pectineus seems an undifferentiated whole partially curving over the pubic-iliac angle (Plate 28a); one can, however, distinguish four separate bundles of muscle cells in tuatara, spreading out across the dorsal surface of the ventral pelvic girdle (Plate 28). Musculus iliopsoas plus Musculus pectineus in the reptiles are considered the "Pubo-ischio-femoralis internus" (Byerly, 1925; Romer, 1922; Gregory and Camp, 1918; Osawa, 1898; Gadow, 1882). There is a modification of the origin of this pectineus muscle in mammals as the ilium changes its shape. The bone points from a posterior dorsal angle in the lizard to the vertical in the alligator, to the parallel and simiparallel and flaired position in the opossum and echidna. There is also a reduction of muscle cell fibers in the insertion. This was to support the weight as the limbs were turned around, as on an axle, and were pulled under the body. This is more pronounced in the opossum; the echidna retains the reptilian thick femur with expanded head which has remained more dorsal compared to the lateral position in the opossum. The carriage in the echidna makes it necessary for strong muscles to support the parts of the legs that maintain the stance.

The action in the opossum is purely flexion; there is a bending, and a turning out of the thigh in the echidna, *i.e.*, pushing back the dirt when digging.

When an animal digs with its fore feet, the function of the hind legs is to provide a resistance to the foreward movement of the body which would normally follow a backward thrust of the fore limb. The muscle which is chiefly responsible for this resistance when the knee is stiffened by its extensors is the iliopsoas, which keeps the hip from extending. . . This muscle and the psoas minor also tend to arch the back; in a low tunnel this brings the back against the roof of the excavation. The hind legs are also of value in passing the detritus backwards; the muscles involved in this action are the same as those responsible for forward progression.[3]

In the advanced reptiles, this iliopsoas-pectineus relationship is found in the crocodile (Romer, 1923) and alligator (Chiasson, 1962). The psoas is mentioned in the turtle (Owens, 1866).

The pubi-ischio-femoralis internus of reptiles has much of the position and functions of the ilio-psoas plus pectineus of mammals, except that it extends further caudad on the inner side of the pelvis.[4]

MUSCULUS PECTINEUS

Didelphis (Plate 1-t: Fig. 3)

Origin: One thick head on the outer base below the marsupial bone (Os epipubis, Fig. 18; Plates 12-e; 12a-f); another thinner mass arises behind the acetabulum and a little below the ilio-pectineal line—the fibers fusing.

Insertion: In two parts—on the border of the femur at the posterior internal ridge, and with a narrow space between, down two thirds the length of the bone.

Tachyglossus (Plate 2-f: Fig. 4)

Origin: Outer border of the marsupial bone (a heterotopic bone in the abdominal wall; *i.e.*, Os epipubis, (Fig. 16; Plates 10a-e; 10b-f) on the pectineal eminence at its juncture to the pubis. This muscle has a flattened spindle shape. It is short and splits in some specimens.

Insertion: Near the Trochanter minor but lower down on the dorsal lateral border of the Corpus femoris.

Sphenodon (Plate 6-p: Figs. 10-p; 11)

Origin: This muscle adds to the intricate mass of the iliopsoas. It comes from the dorsal surface of the ventral pelvic plate—probably the fourth bundle part is the one emerging from the epipubis.

Insertion: The fibers meld with the iliopsoas at the obtuse mesial side of the pubio-iliac angle, above the acetabulum—inserting nearer the femur body than the rest of the muscle.

In *Didelphis* this muscle can sometimes be found in contact with the Ligamentum inguinale (from the ilium). It would be easy to consider it an adductor as it draws the thigh toward the middle line of the body and turns the femur outward. This is also true in *Tachyglossus*. Here it shows one or two flattened strips, shorter than in *Didelphis*. The Musculus pectineus is found split even more in some specimens. In *Sphenodon*, Gregory and Camp point out,

The pectineus has much of the position and function of the ilio-psoas plus pectineus of mammals, except that it extends further caudad on the inner side of the pelvis.[5]

It is always a dorsal muscle.

In *Sphenodon*, Romer (1922, 1942) considers

[3]Elftman, H. 1929. Adaptations of the marsupial pelvis. Bull. Amer. Mus. Nat. Hist., 58:216. Courtesy of the American Museum of Natural History.

[5]Gregory, W., and C. Camp. 1918. Studies in comparative myology and osteology. Bull. Amer. Mus. Nat. Hist., 38:486. Courtesy of the American Museum of Natural History.

[4]Gregory, W., and C. Camp. 1918. Studies in comparative myology and osteology. Bull. Amer. Mus. Nat. Hist., 38:486. Courtesy of the American Museum of Natural History.

CHART OF PELVIC MUSCULATURE

MUSCLES	TUATARA	ECHIDNA	OPOSSUM	INNERVATIONS
	SPHENODON—RHYNCHOCEPHALIAN	TACHYGLOSSUS—MONOTREME	DIDELPHIS—MARSUPIAL	
M. iliopsoas	o. dorsal surface of ventral pelvic plate; membrana obturatoria	o. ilium; lumbar sacral vertebrae	o. ilium, lumbar, vertebrae sacral	Plexus lumbalis
	i. presacral vertebrae; medial femur	i. femur neck	i. femur trochanter	
M. pectineus	o. dorsal surface of ventral pelvic plate	o. outer marsupial bone	o. below marsupial bone; behind acetabulum	Plexus lumbalis
	i. femur body	i. lateral femur	i. along femur	
Mm. glutei	o. lateral ilium	o. sacral, dorsal vertebrae; ilium	o. surface of ilium; fascia	Plexus sacralis (Tachyglossus — Plexus lumbo-sacralis)
	i. ventral femur	i. femur body	i. femur trochanter and body	
M. femorococcygeus	o. caudal vertebrae wing	o. sacral and caudal vertebrae connective tissue	o. fascia rear caudal vertebrae (3rd)	Plexus sacralis
	i. femur neck	i. lateral tibia	i. caudal lateral tibia	
M. sartorius	o. prepubic tubercle	o. iliopectineal eminence	o. anterior ilium; ligamentum, inguinale	Plexus lumbalis
	i. fascia tibial tuberosity	i. femur condyle; Knee joint	i. ligamentum patella	
M. rectus femoris	o. lateral prepubis ilium	o. bottom of iliac shaft	o. ilio pectineal eminence	Plexus lumbo-sacralis (Didelphis-Plexus lumbalis)
	i. mesio-lateral tibia head	i. ligamentum patella tibia head	i. ligamentum patella	
Mm. Vasti	o. lateral femur	o. dorsal femur	o. femur neck	Plexus lumbo-sacralis (Didelphis — Plexus lumbalis)
	i. fascia tibial tuberosity	i. tendon of the patella	i. tendon to patella	
M. gracilis	o. symphysis and ligament of pubio-ischiadicus	o. ischium, pelvic symphysis: marsupial bone: horizontal pubis	o. ventral pelvic symphysis; horizontal pubis	Plexus lumbo-sacralis (Didelphis — Plexus lumbalis)
	i. tibial tuberosity and crest	i. posterior tibia shaft	i. tibial crest; mesial shaft	
Mm. adductors	o. fascia of prepubis; ligamentum pubioischiadicus; pubic tuberosity	o. epipubic bone; pelvis symphysis; ischiadic arch	o. ilio pectineal eminence; horizontal pubis; pubic symphysis; ischiadic arch	Plexus lumbo-sacralis (Didelphis — Plexus lumbalis)
	i. under femur lateral tibia	i. medial femur	i. ventral femur body	
M. biceps femoris	o. sacral vertebra	o. pelvic arch; ischial tuberosity	o. tuberosity of ischium	Plexus sacralis (Sphenodon — Plexus lumbalis)
	i. lateral fibula	i. fascia along and around tibia shaft	i lateral tibia; fascia of fibula	
M. semimembranosus	o. fascia of 1st caudal vertebrae; ischiadic arch	o. tuberosity of ischium	o. caudal ischium; tuberosity of ischium	Plexus lumbo — sacralis (Didelphis — Plexus sacralis
	i. femur trochanter	i. mesial tibial crest	i. tuberosity of tibia	
M. semitendinosus	o. tuberosity of ischium; arch	o. with semimembranosus; tuberosity of ischium	o. tuberosity of ischium	Plexus lumbo — sacralis (Didelphis — Plexus sacralis)
	i. lateral tibia	i. dorsal tibia	i. body of tibia	

COMPILED FROM OBSERVATIONS AND TABLES OF PELVIC MUSCLE PATTERNS this the "Pubo-ischio-femoralis-internus," as does Byerly (1925). Romer (1923) found only two parts in the lizard. Osawa (1898) calls it "Pubi-ischio-trochantericus-internus." Gadow (1882) gives it the same name but makes four parts out of it, *i.e.*, three distinctive divisions of "Pubo-ischio femoralis internus," and a posterior portion.

In primitive reptiles, the so called pubioischio-femoralis internus was a large muscle mass. Part of it moved up on the ilium and part remained in evolution; this piece, the pectineus in the mammals, with the Musculus iliopsoas makes up the reptilian homolog (Romer, 1922).[6] Its function generally is the same in the reptile, echidna and the opossum. Noble and Noble (1939) and Ashley (1955) show only the internal iliacus as its homolog in the turtle. Owens (1866) demonstrates a pectineus in *Chelonia.* Romer (1923) found two parts of this pubioischiofemoralis internus in the crocodile. In this more advanced reptile, the origin had been reduced a bit, and had moved more dorsally—a breaking up of the primitive bundle seen in *Sphenodon.* Romer (1942) worked on the lizard and verified the Musculus pectineus coming from the internal surface of the girdle. Gadow (1882) and Chiasson (1962) show in the alligator a pubioischiofemoralis internus.

Such similarities in all orders are reinforced by the fact that the Plexus lumbalis and its twigs innervate the pectineus. Specifically in *Sphenodon*, Nervus obturatorius and Nervus femoralis are entailed. The latter, in *Tachyglossus*, has two branches at the neuro-muscular juncture. Only one branch from Nervus femoralis was found in any *Didelphis*.

[6] Romer, A. 1922. The locomotor apparatus of certain primitive and mammal-like reptiles. Bull. Amer. Mus. Nat. Hist.

2. Gluteal Group

These muscles collectively are innervated by the branches of the femoral and crural nerves (Plexus lumbosacralis).

MUSCULI GLUTEI

Didelphis (Plate 3-i,j,g: Fig. 5)

Origin: Top of the ilium along the dorsal fascia of the lumbar, the sacral, and caudal fascia, the iliac crest, the gluteal surface of the ilium, and on its lower edge.

Insertion: On the Trochanter major and caudal side of the femur.

Tachyglossus (Plate 4-c,d,e,j,t,z: Figs. 6; 7)

Origin: On the spines of the sacral and dorsal vertebrae, the sacral vertebrae, along the wing and body of the ilium to just above the acetabulum, and from the lumbo-dorsal fascia.

Insertion: On the distal edge of the Trochanter major and posteriorly and anteriorly along the length of the body of the femur.

Sphenodon (Plate 6-i: Figs. 10-i; 11)

Origin: Below the Musculus biceps femoris on the lateral surface of the ilium together with Musculus tensor fascia femoris.

Insertion: Anteriorly on the ventral proximal half of the shaft of the femur along the gluteal ridge.

There are three divisions of this muscle in the opossum, and there are degrees of depth and surface parts. The long fibers which originate in the iliac crest and ventral iliac spine and tuberosity circle and wrap themselves around a part of the body of the ilium. The rest of the divisions fan out from the gluteal surface of the ilium on a broadening thoraco-lumbar connective tissue. This extends from the vertebral column to just about opposite the tuberosity of the ischium, and laterally to the Musculus longissimus lumborium. Some fibers go into the great trochanter, and extend two thirds of the way along the body of the femur as far as the lateral lip of the shaft. These muscles in the opossum are the principal inverters of the thigh in aboreal life.

Summarily, there are about six musculi glutei bundles in the echidna contrasted with the three in the opossum, and the one in tuatara.

On the echidna (Plate 4) they comprise what I call a small pars caudalis, t, of the larger bundle, j, a gluteus superficialis, c, and three deep groups, d, e, and z. They arise from the whole outer surface and the gluteal inner wing of the ilium with no fascia. Some pieces are from the anterior border of this shaft and on its ventral surface; furthermore, portions of fibers are seen on sacral and coccygeal fascia and vertebrae. This detailed description explains the large mass as an adaptation for digging. All muscle bundles are very thick, and the cells run deep—one mass overlapping another. The muscles vary in the number of their parts in the various orders; generally, in some animals they can be classified in layers, but in the primitive mammals what is seen are superficial and various deep arrangements which differ in position in the subclasses, with the branching nerves following the splitting tissue.

In the tuatara this one muscle bulk arises only on the side of the ilium and is covered by the Musculus biceps femoris. It can be seen in *Didelphis* (Plate 3: Fig. 5) inserting on the body of the femur—there appears to be some trochanteric attachment in the marsupials studied. In *Sphenodon*, Byerly (1925), and Gregory and Camp (1918), Romer (1922), and Osawa (1898) designate this Musculus gluteus as Musculus iliofemoralis. Gadow (1882) in primitive reptiles, considers this part of the "caudi-ilio-femoralis." Its small fusiform shape and the fascia which binds it to the ilium show its primitive character, and accompanying lack of use in tuatara. It is innervated by the Nervus femoralis branches, specifically Nervi obturatorius and ischiadicus in the echidna (*Tachyglossus*), the latter nerve alone in *Sphenodon* and *Didelphis*.

In the turtle, Ashley (1955), Owen (1866) and Noble and Noble (1939) have retained the word gluteus for this muscle. Here it begins on the head of the femur and the sacral vertebrae. Chiasson (1962) calls it ilio-femoralis or caudi-ilio-femoralis and he has it originating from the iliac dorsolateral area in the alligator. In the crocodile, Romer (1923) has given the name iliofemoralis to it, and here its beginning is on the head of the ilium, on its posterior lateral side to just above the acetabulum. Romer (1970) called it iliofemoralis in the lizard alone. To Gadow (1882) in the alligator it was the caudi-ilio-femoralis, but Gregory and Camp (1918) considered that part came from the ilio-femoralis and the rest from the ilio-tibialis. The general opinion seems to be that the Musculi glutei are the extensions of the reptilian ilio-femoralis, which has been split, expanded, thickened, lengthened, or shortened, progressively increasing, specializing, restricting, and increasing independence of muscle masses. These bundles, as they change their reptilian type in the lizard, become distinctive and specialized in the turtle with no lumbar vertebrae as such (Kent, 1965), elongated in *Didelphis*, the opossum, and solidly condensed in *Tachyglossus*, the echidna, basically occurring due to the change of the direction of the ilium from a dorsoanterior direction to one which lies in an anterior direction. Tuatara has maintained the position of origin only on the ilium, which is slightly tilted backwards; the general area of insertion on the femur is nearly the same.

MUSCULUS FEMOROCOCCYGEUS (FEMOROCAUDALIS)

Didelphis (Plate 3-d: Fig. 5)

Origin: On the fascia near the third caudal vertebra.

Insertion: On the caudal, lateral side of the femur.

Tachyglossus (Plate 4-a: Fig. 6)

Origin: On connective tissue of the sacral and coccygeal (caudal) vertebrae.

Insertion: Distally and laterally on the tibia.

Sphenodon (Plate 5-q ?: Figs. 8-q; 9)

Origin: On caudal vertebra wing.

Insertion: Proximally, on neck of the femur.

Romer (1922) does not show this opossum muscle. Since it is innervated by a gluteal nerve, it leads to the conclusion that he considers it a fourth gluteus. In our specimens studied, its existence was uncertain. It is found in the work of Coues (1872); Elftman (1929) describes it as an extender of the hip, and adductor of the thigh.

> In arboreal types the muscle may be of use in pulling the hind limbs toward the tail. It may also be used as a lateral flexor of the tail.[7]

It is prominent in Osgood's (1921) *Caenolestes*; Barbour (1963) calls it agitatorcaudae and says it can not be separated very well from a gluteus in marsupials. Unless this is a gluteus in the echidna, femorococcygeus would not be an exact descriptive term for these fibers, and so it would be considered lacking. The muscle shown on Plate 4-a could be a fragmented slip of Musculus cruroccygeus.

[7]Elftman, H. 1929. Adaptations of the marsupial pelvis. Bull. Amer. Mus. Nat. Hist., 58:205. Courtesy of the American Museum of Natural History.

In *Sphenodon*, caudi-femoralis (Byerly, 1925) was noted on the proximal anterior surface of the femur to be innervated by a spinal and caudal nerve. Gadow (1882) considers this muscle to be in two parts among the reptiles; Osawa (1898) and Romer (1923) have a synonym in coccygeofemoralis longus and brevis.

A muscle, caudi-iliofemoralis (Gregory and Camp, 1918) is the posterior part of the iliotibialis externus in the reptile. There is a deep muscle in the tuatara studied (Plate 5-q: Figs. 10; 11) which could be so named, but it is attached by tendons to the Musculi semimembranosus-semitendinosus. (These two together are the flexor tibialis, internus and externus, of reptilian muscle nomenclature).

In the cat, Reighard and Jennings (1963) call this the caudofemoralis. A true caudofemoralis is a hamstring muscle and is innervated by the hamstring branch of the tibial nerve. It does not occur in the *Homo sapiens*, (or is it vestigeal, or Musculus piriformis?).

It seems that a Musculus femorococcygeus is found in some marsupials and is lacking in the echidna, unless it can be considered one of the glutei. In tuatara, an extensive segmented caudal muscular system is found, very primitively developed, lying among the thirty or so post-sacral (caudal) vertebrae. Chevron bones abound.

3. Musculus Sartorius

Didelphis (Plates 1-a; 3-h: Figs. 3; 5)

Origin: From the anterior external border of the ilium and at the extreme upper end of the Ligamentum inguinale (pouparti).

Insertion: Anterior and medial edge of the Ligamenta patellae.

Tachyglossus (Plate 2-d: Fig. 4)

Origin: From a low inception on the enlarged

ilio-pectineal eminence.

Insertion: On the median side of the joint of the knee to fascia above the tibia; on the side of Ligamentum cruciatum and below the median internal condyle of the femur.

Sphenodon (Plates 5-h; 6-w: Figs. 8-h; 9)

Origin: By fascia from tubercle of the prepubis.

Insertion: By more extensive fascia blending with other connective tissue over the knee to the proximal tuberositatis tibiae.

In *Didelphis*, this muscle is a narrow strip of fibers which get thicker as they extend caudally; this is definitely an isolated mass which brings the thigh forward as the opossum walks and lifts the leg. It acts to bend the femur and extend the limb. In *Tachyglossus* this large muscle, together with the Musculus iliacus portion of Musculus iliopsoas, supposedly is an adaptation which accounts for the broad ilia. These sturdy bones keep the echidna from moving forward as it digs with its feet. The shape of this same muscle in the echidna is flattened and triangular. It helps rotate, bend, and turn the leg on its axis. In *Sphenodon*, a "bottled-shaped" muscle originates on fascia and tendons, its broad end inserting with much connective tissue into the tibia. Here the muscle is poorly developed. This demonstrates lack of use (due to diminished function) even though it helps extend the leg to some degree; it does not aid in rotating the femur.

In the marsupial, the Musculus sartorius is supplied by branches of the femoral nerve. In the monotreme, there are slight variations, but the same innervations persist. The innervations in the rhynchocephalian neuromuscular juncture come from the Plexus lumbalis (the Nervus femoralis and Nervus obturatorius). All three genera have fairly similar nerve fiber relationships.

When there is homologizing and synomymizing to be found in the literature concerning reptilian muscle cells and those of primitive mammals, every paper has different points of of view.

> . . . the ambiens (of reptiles) seems to be the homologue of the monotreme sartorius, which is clearly that of higher mammals.[8]

> . . . The "sartorius" is not, of course, the mammalian muscle of that name (which is the sauropsid ambiens) nor can it equal the reptilian pubo-tibialis, which is a ventral muscle.[9]

Romer (1970) often questions the homology. In the opossum, the origin is not associated very closely with the quadriceps. The psoas in the echidna inserts on the pelvis in such a way that it is also the origin of the sartorius; the sartorius could be considered an extension of the Musculus psoas. Gadow (1882) and Osawa (1898) consider ambiens in the primitive reptile, *Sphenodon*, the "pubi-tibialis (posticus)", which originates on the tubercle of the pubis and inserts on the lateral or medial prominence of the tibia. Here it is innervated by the Nervus femoralis and Nervus obturatorius. But Gadow (1882) does not believe that this muscle is the sartorius in mammals.

> . . . the sartorius of mammals has been derived from the ilio-tibialis internus, which may have been present in primitive reptiles.[10]

Byerly (1925) considers ambiens in *Sphenodon* to

---

[8]Romer, A. 1922. The locomotor apparatus of certain primitive and mammal-like reptiles. Bull. Amer. Mus. Nat. Hist., 46:563. Courtesy of the American Museum of Natural History.

[9]Romer, A. 1923. Crocodilian pelvic muscles and their avian and reptilian ancestry. Bull. Amer. Mus. Nat. Hist., 48:546. Courtesy of the American Museum of Natural History.

[10]Gregory, W., and C. Camp. 1918. Studies in comparative myology and osteology, 38:492. Bull. Amer. Mus. Nat. Hist. Courtesy of the American Museum of Natural History.

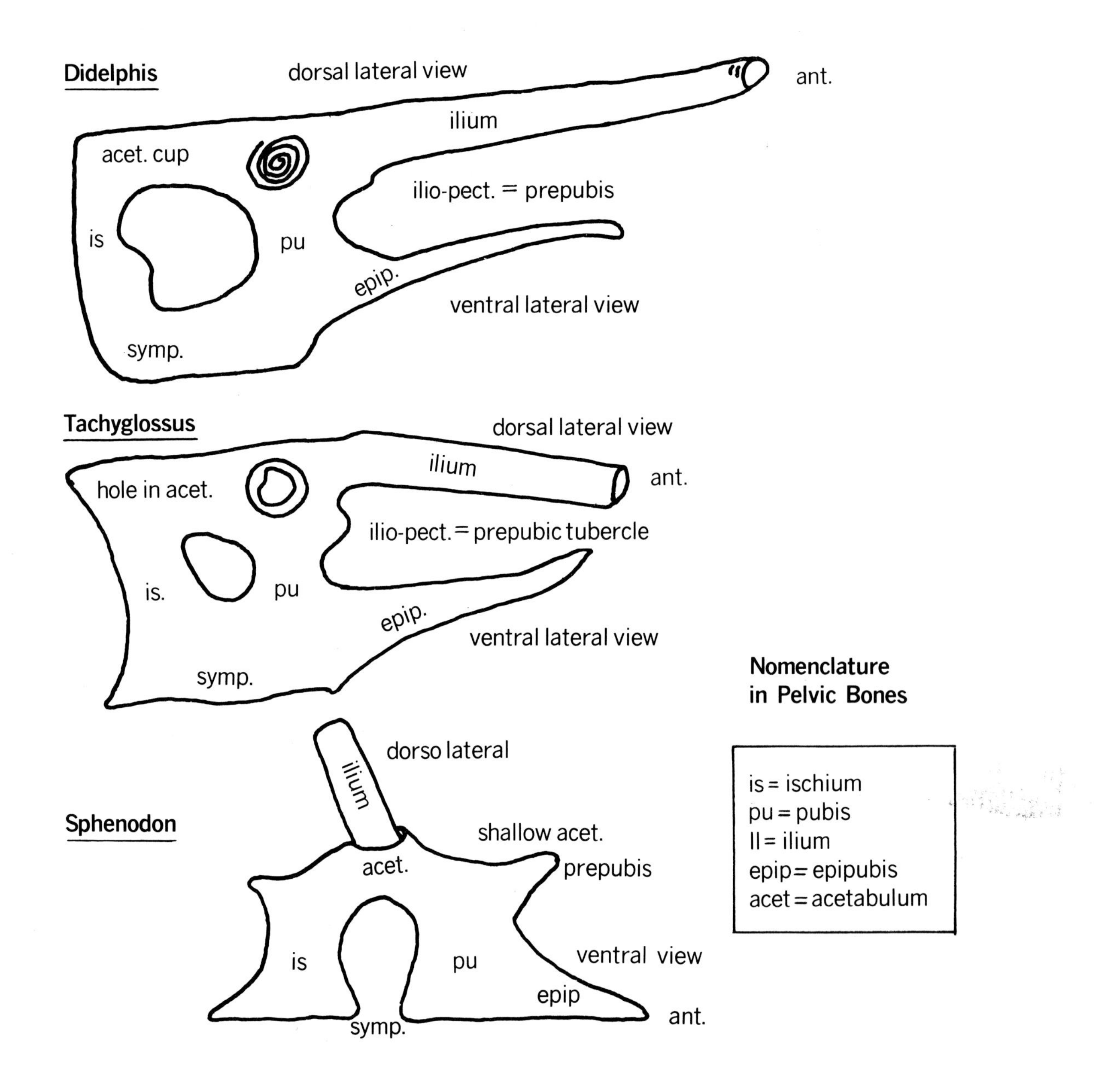

Nomenclature in Pelvic Bones

is = ischium
pu = pubis
Il = ilium
epip = epipubis
acet = acetabulum

In all cases, the ossa innominata (iliums) have a juncture with the sacrum; they form the pelvic arch.

TABLES OF PELVIC MUSCLE PATTERNS

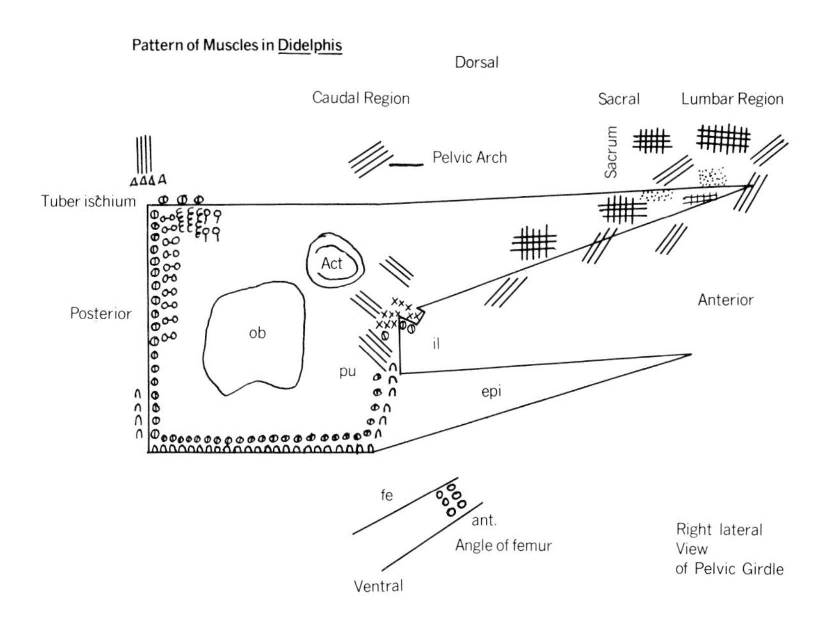

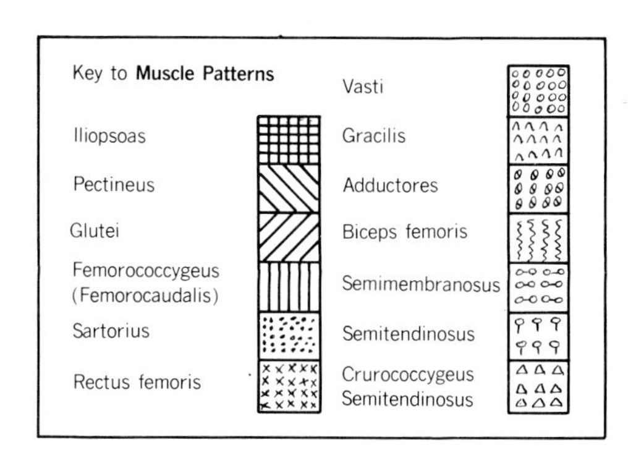

(Key for "Pattern of Muscles")

act = acetabulum
il = ilium
is = ischium
pu = pubis
pp = prepubis
epi = epipubis
ob = obturator
fe = femur
ant = anterior

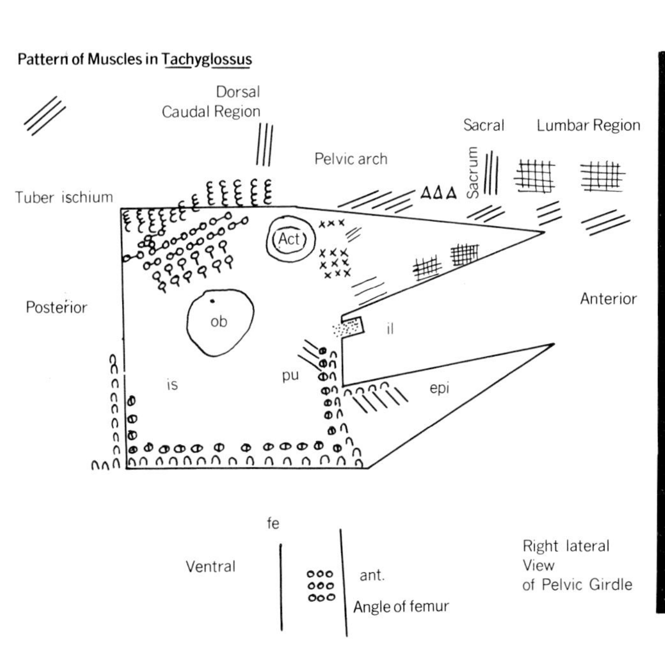

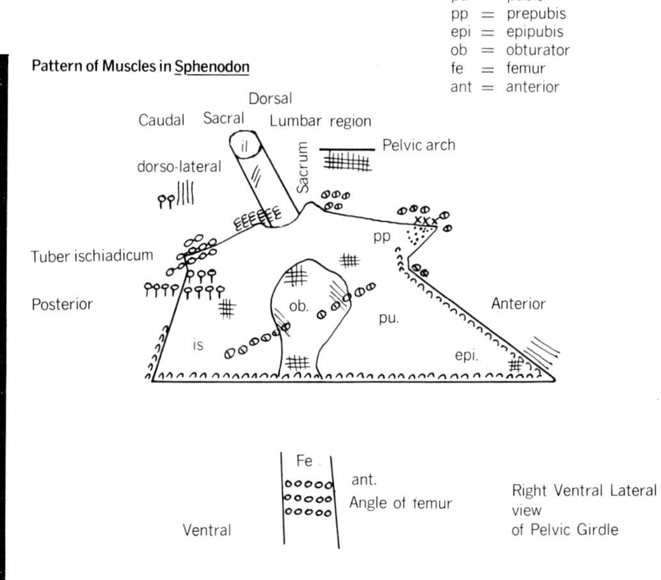

blend with Musculus ilio-tibialis (Musculus rectus femoris) and Musculus ischio-femoralis (Musculus adductor) in continuous fascia, lying on the anterior top of the tibia, and on the anterior surface of the femur. These three make up a Musculus triceps femoris.

To sum up, the consensus is that this muscle, ilio-tibialis internus, is the general reptilian condition, found in the alligator and the crocodile (Gregory and Camp, 1918; Romer, 1923). This muscle has two heads in the lizards and is triple-headed in the crocodile arising from the dorsal border of the ilium. It has a double innervation in the reptiles, *i.e.,* from the crural and sacral plexus. In the alligator, the Musculus ambiens as referred to by Chiasson (1962) may be the Musculus iliotibialis. Noble and Noble (1939) find a Musculus sartorius in the turtle that begins in the pelvic ligament, originating by a tendon, and inserting on the proximal end of the tibia, supposedly extending the leg. The Musculus attahens pelvim (Ashley, 1955) in the turtle, originating on the pectineal or lateral pubic process may be a homolog of Musculus sartorius. The insertion of this muscle is on the plastron posteriorly, and is considered a muscle fixing the pelvic girdle. The Musculus triceps femoris is an extender of of the leg in this instance in the turtle.

The general impression is that in the conservative, non-evolving animals the Musculus sartorius has been kept. Its homolog in many research papers, on various animals of all kinds, is very controversial.

4. Quadriceps Femoris Group

These muscles are innervated generally by branches of the femoral and crural nerves (Plexus lumbo-sacralis).

MUSCULUS RECTUS FEMORIS

Didelphis (Plates 1-b; 3-o: Figs. 3; 5)

Origin: From the iliopectineal eminence on a dorso-lateral ridge above the acetabulum with two united heads, a small and a large one.

Insertion: With a blending of the two components on a small tendon, into the Ligamentum patella with the Musculi vasti.

Tachyglossus (Plates 2-b; 4-w: Figs. 4; 6)

Origin: By two blended heads on a stout flattened tendon from the projection at the bottom of the iliac shaft, above the circular cap over the lunate surface of the acetabulum.

Insertion: One branches into the Ligamentum patella accompanying the Musculi vasti; the other goes over the knee into the head of the tibia.

Sphenodon (Plate 6-l: Figs. 10-l; 11)

Origin: On the lateral surface of the prepubis, the fibers by heads on short tendons; the whole mass slanting downward: iliac process.

Insertion: Into the mesio-lateral head of the tibia on a tendon closely associated with a Musculus vastus (Musculus femor tibialis).

In *Didelphis* this muscle has no femoral relationship; along some of its length it hangs over the vasti. It is slightly cone-shaped and cylindrical in cross section; there is a merging of different sized fibers. In *Tachyglossus* this muscle is unrelated, separate and distinct from the vasti, as in all lower animals where accessory tendons abound. This is true in *Sphenodon*, also. It is pyramidally shaped and separates or divides into insertions. In *Sphenodon*, too, some investigators consider it a Musculus gluteus (*i.e.*, Westling, 1889, etc.) and most of these give it the reptilian name Musculus ilio-tibialis.

In *Tachyglossus*, the heads are one as in the

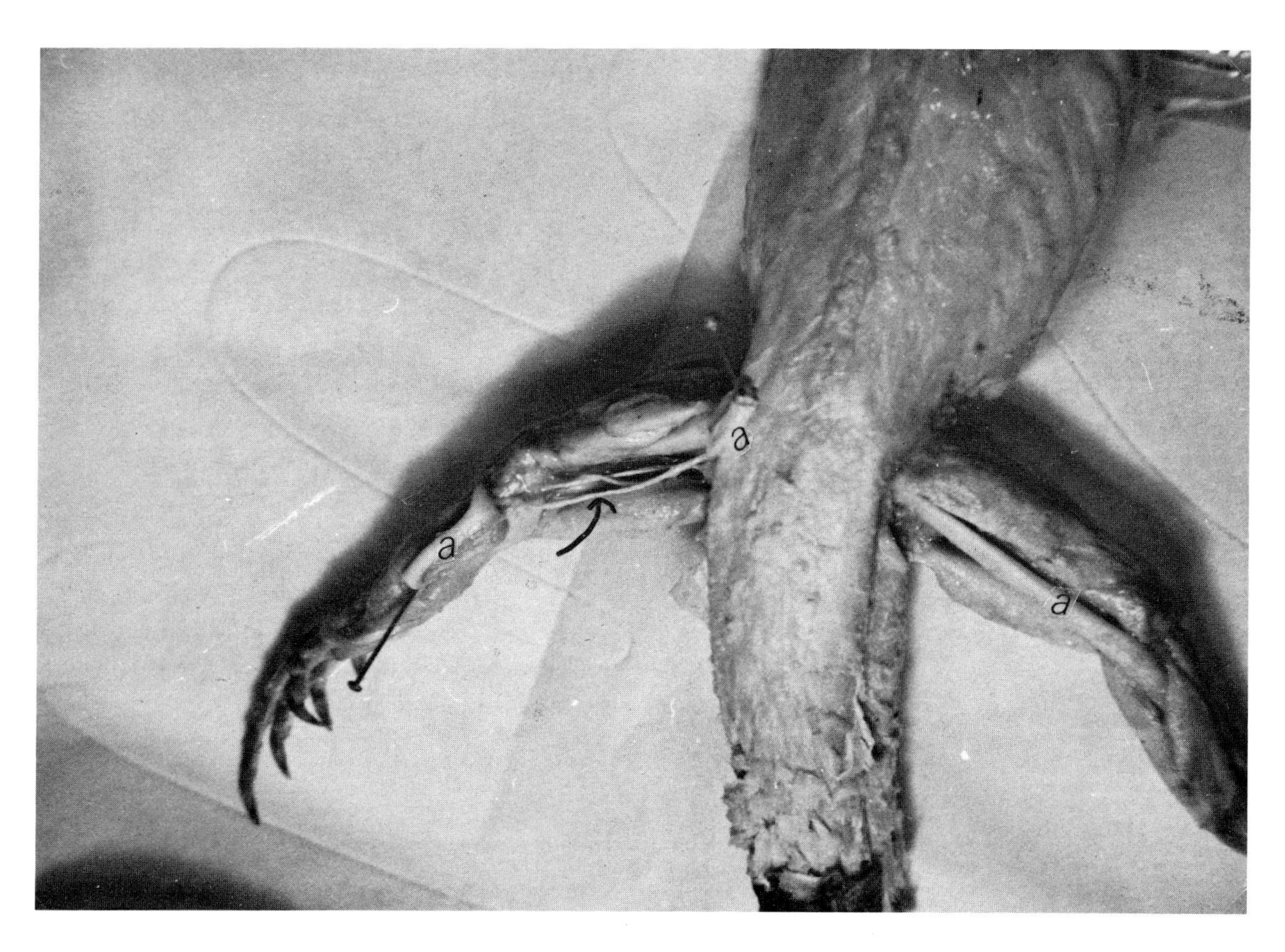

This plate shows the PLEXUS LUMBOSACRALES at arrow. Note Musculus biceps femoris at a. Dorsal view of Sphenodon.

opossum, *Didelphis*; however, the muscle arises at the base of the ilium, and is seen originating more prominently above the acetabulum. The insertions are the same in all genera except in this echidna where a few fibers overlap into the tibia. *Sphenodon* has no ilial origin but extends its fibers from the pelvic plate by primitive tendons. The fascicular bundle is prominently included in the tibial tendon of this tuatara's Musculus rectus femoris. In all three primitive organisms, the Musculi vasti are below the muscle masses. The muscle in all cases is a massive extensor of the thigh.

In the reptiles, in general, the Musculus rectus femoris, as part of the quadratus femoris in primitive mammals, is likened to the iliotibialis (Romer, 1942; Hyman, 1944). Gadow (1882) and Osawa (1898) in *Sphenodon* called it ambiens or pubio-tibialis. To them, Musculus ilio-tibialis was a different muscle, what Westling (1889), as mentioned above, considred a Musculus gluteus maximus in the monotreme. Gregory and Camp (1918) thought it probably derived from the ambiens. With this matching in structure, it originates in generalized reptiles near the base of the pubic tuberculum and inserts with a tendon of the Musculus femorotibialis on the anterior surface of the head of the tibia. Gadow (1882) notes two parts, while Romer (1923) finds three in Crocodilia. Noble and Noble (1939) in the turtle hangs the Musculus rectus femoris by a tendon on the ilium, and puts its insertion on the metatarsus of the last digit of the hind foot. Owens (1866) finds a Musculus rectus femoris in the turtle. This reptilian muscle seems changed by the upgrowth of the gluteals in the therapsid mammals. Romer (1922)[11] suggests that the gluteals have shoved the reptilian Musculus ilio-tibialis anteriorly and ventrally as the Musculus rectus femoris; he explains that this crowding out has limited it to the anterior part of the ilium. In the three conservative animals studied, all innervations were traced to the offshoots of the Nervus lumbosacralis. In *Didelphis*, only the Nervus femoralis goes to the Musculus rectus femoris. In *Sphenodon* and *Tachyglossus*, the twigs of the Plexus sacralis or Nervus femoralis are found at the nerve-muscle juncture. This muscle bundle in the true mammals does not have the crural and sacral nerves to it, that are found in the Reptilia.

MUSCULI VASTI

Didelphis (Plates 1-c,d; 3-f,v: Figs. 3, 5)

Origin: On the trochanter major on one side; an inner and an outer thin lower mass surrounds the shaft of the femur at the neck.

Insertion: With the rectus femoris on the patella by a small wide tendon.

[11] Romer, A. 1922. The locomotor apparatus of certain primitive and mammal-like reptiles. Bull. Amer. Mus. of Nat. Hist., 46:527.

Tachyglossus (Plates 2-a,p; 4-g,h,i; 4A: Figs. 4; 6; 7)

Origin: In a plump fleshy bulk from the broad anterior dorsal surface of the femur at the point where lie the Musculus iliopsoas and the Musculi glutei.

Insertion: On a tendon which leads the Musculi vasti into the patella.

Sphenodon (Plates 5-e; 6-j,k,m,n: Figs. 8-e; 9; 10-j,k,m,n; 11)

Origin: Enraps the femur (from the insertion of iliopsoas and pectineus, or the pubio-ischio femoralis internus) spreading fan-shaped on the medial and lateral sides.

Insertion: On the fascia at the edge of the tibial tuberosity.

In the opossum, *Didelphis*, these muscles blend their whole length to a certain extent and are inseparable; they lie entirely on the femur. In the echidna, *Tachyglossus*, there appear to be five separate bundles that lie entirely on the thigh. Interrupted by the Musculus rectus femoris anteriorly, they collectively encircle this part of the femur. In the tuatara, *Sphenodon*, this muscle is called Musculus femoro-tibialis by some (Byerly, 1925; Gregory and Camp, 1918; Romer, 1942; Hyman, 1944). Osawa (1898) and Gadow (1882) consider its name to be Extensor Triceps. Again the great amount of fascia at its insertion in this rhynchocephalian, *Sphenodon*, seems to show transitional adaptation.

The origin in *Didelphis* shows two parts near its head. *Tachyglossus* Musculi vasti arise as one piece near other muscles. In *Sphenodon*, this mass more extensively surrounds the femur.

Insertion here is on the fascia of a leg bone and the tibial crest, while in the echidna and the opossum there are only patella attachments.

The femoral nerve gives off branches into the marsupial *Didelphis'* Musculi vasti, and this is true in the monotreme, *Tachyglossus*. Rhyncho-

TABLES OF NERVE PATTERNS TO SPECIFIC MUSCLES (Compiled from Research and Dissected Specimens)

Sphenodon

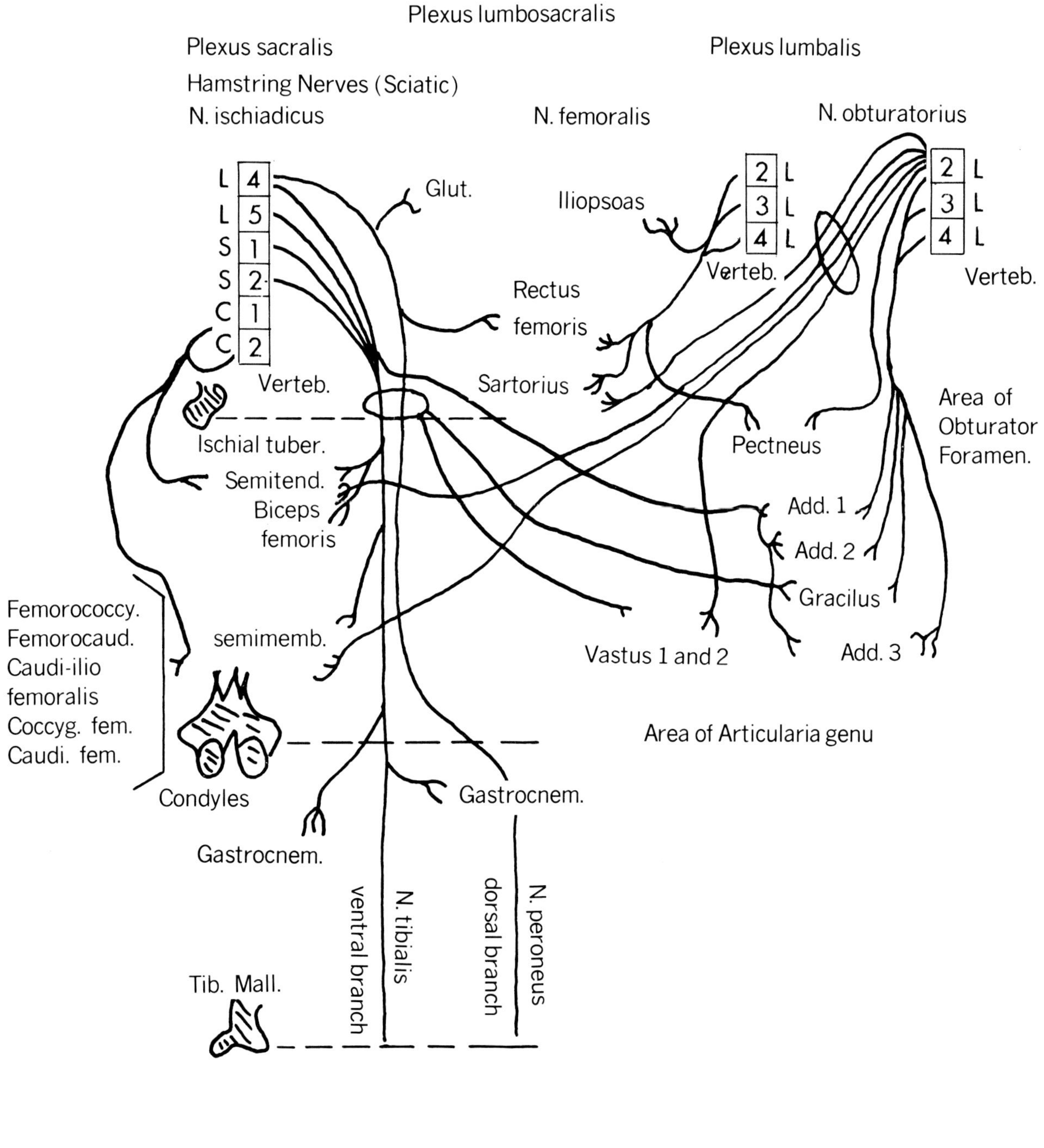

Plexus lumbosacralis

Plexus sacralis

Hamstring Nerves (Sciatic)

N. ischiadicus

N. femoralis

Plexus lumbalis

N. obturatorius

L 4
L 5
S 1
S 2
C 1
C 2

Glut.

Iliopsoas

2 L
3 L
4 L

2 L
3 L
4 L

Verteb.

Verteb.

Rectus femoris

Verteb.

Sartorius

Area of Obturator Foramen.

Ischial tuber.

Pectneus

Semitend.

Biceps femoris

Add. 1

Add. 2

Gracilus

Femorococcy.
Femorocaud.
Caudi-ilio femoralis
Coccyg. fem.
Caudi. fem.

semimemb.

Vastus 1 and 2

Add. 3

Area of Articularia genu

Condyles

Gastrocnem.

Gastrocnem.

ventral branch

N. tibialis

dorsal branch

N. peroneus

Tib. Mall.

NERVE PATTERN Didelphis

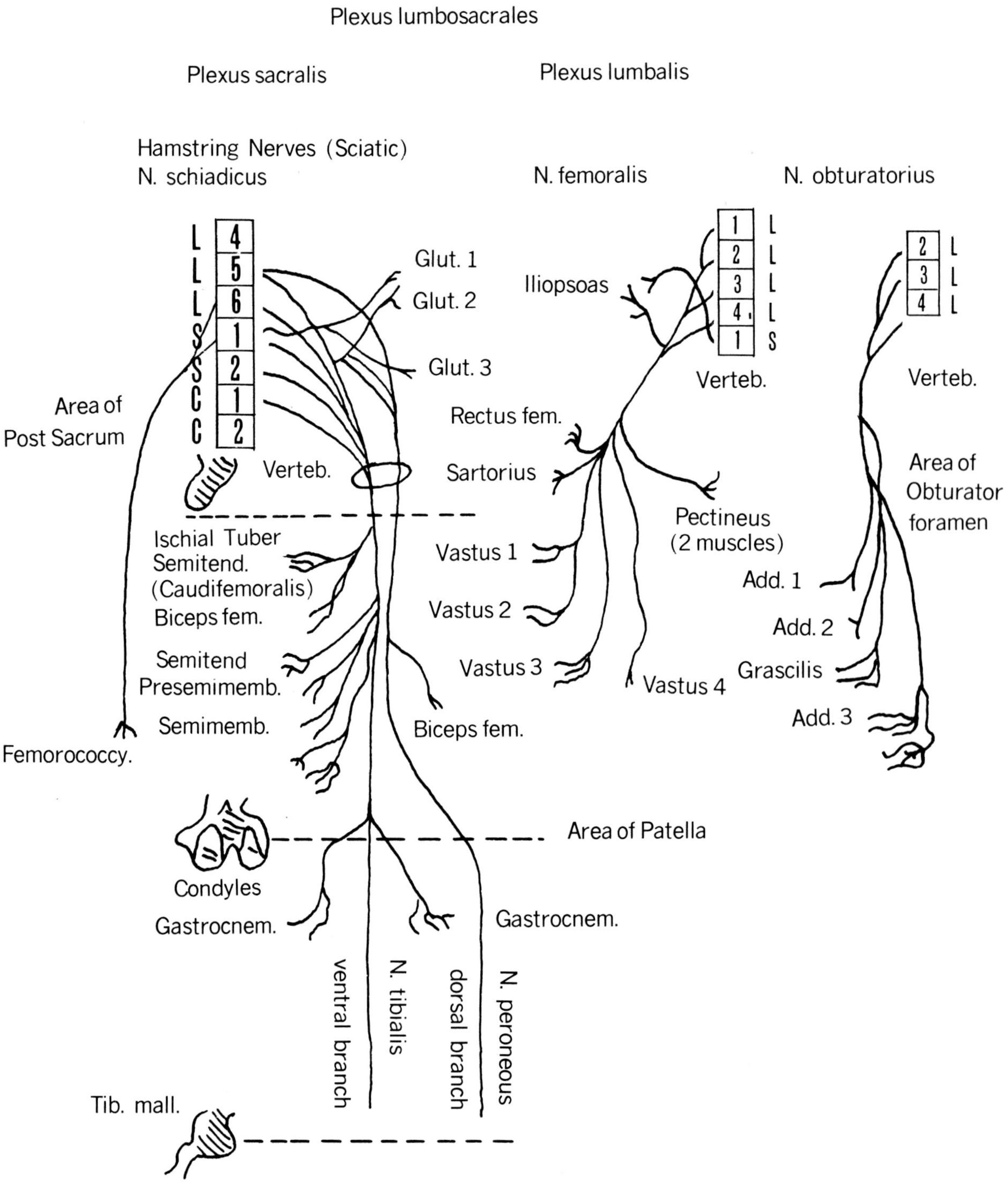

Plexus lumbosacrales

Plexus sacralis

Hamstring Nerves (Sciatic)
N. schiadicus

L 4
L 5
L 6
S 1
S 2
C 1
C 2

Glut. 1
Glut. 2
Glut. 3
Area of Post Sacrum
Verteb.
Ischial Tuber
Semitend.
(Caudifemoralis)
Biceps fem.
Semitend
Presemimemb.
Semimemb.
Biceps fem.
Femorococcy.
Area of Patella
Condyles
Gastrocnem.
Gastrocnem.
ventral branch
N. tibialis
dorsal branch
N. peroneous
Tib. mall.

Plexus lumbalis

N. femoralis

1 L
2 L
3 L
4 L
1 S

Iliopsoas
Verteb.
Rectus fem.
Sartorius
Pectineus
(2 muscles)
Vastus 1
Vastus 2
Vastus 3
Vastus 4

N. obturatorius

2 L
3 L
4 L

Verteb.
Area of Obturator foramen
Add. 1
Add. 2
Grascilis
Add. 3

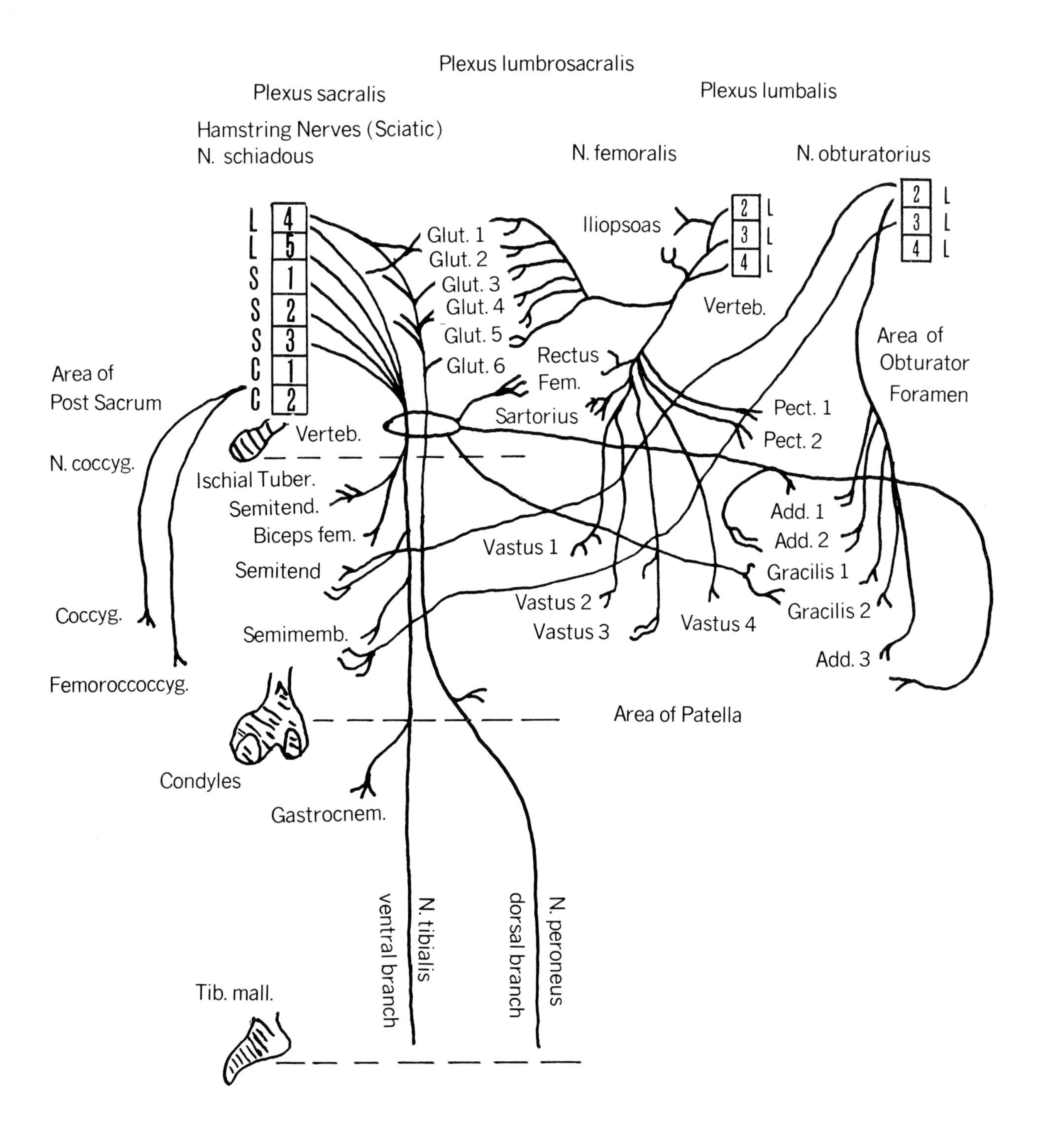

Plexus lumbrosacralis
Plexus sacralis
Plexus lumbalis
Hamstring Nerves (Sciatic)
N. schiadous
N. femoralis
N. obturatorius
L 4
L 5
S 1
S 2
S 3
C 1
C 2
2 L
3 L
4 L
Glut. 1
Glut. 2
Glut. 3
Glut. 4
Glut. 5
Glut. 6
Iliopsoas
Verteb.
Rectus
Fem.
Sartorius
Area of
Obturator
Foramen
Area of
Post Sacrum
Pect. 1
Pect. 2
Verteb.
N. coccyg.
Ischial Tuber.
Semitend.
Biceps fem.
Semitend
Semimemb.
Add. 1
Add. 2
Vastus 1
Gracilis 1
Gracilis 2
Vastus 2
Vastus 3
Vastus 4
Add. 3
Coccyg.
Femoroccoccyg.
Area of Patella
Condyles
Gastrocnem.
ventral branch
N. tibialis
dorsal branch
N. peroneus
Tib. mall.

cephalian, *Sphenodon*, has definite twigs from Nervus obturatorius and the Nervus ischiadicus. The same action can be performed in all genera, with adduction of the thigh and extension of the leg. In the turtle, there is a segregation of muscle fibers, and this is typically so because of the advent of the encasing carapace and plastron; here also is the elimination trend at work in evolution in a specialized organism. Ashley (1955) and Noble and Noble (1939) in the turtle report a Musculus vastus internus coming laterally from the pubic process of the girdle and edge of the femur, inserting with the Musculus satorius on the tibia. In the alligator, Chiasson (1962) shows the Musculus femoro-tibialis internus arising by two fleshy heads on the anterior and posterior surfaces of the femur with the Musculus iliofemoralis (gluteus), and ending by a tendon with other muscles over the knee to the tibia; the shank is extended. In the monotreme, *Tachyglossus*, Gregory and Camp (1918) make this part of Musculus quadriceps femoris.

The Musculi vasti are homologized by Romer (1922) in typical reptiles to the femoro-tibialis, and are considered by him to be masses of muscle first developed in the reptiles. In Romer's (1923) investigation of the crocodilian myology, splitting of the bulk had taken place, and attachments to he femur were by additional heads—the total numbering four. Gadow (1882) made two muscles out of femoro-tibialis, *i.e.*, an inner and outer head in reptiles.

The same general relationships exist in conservative animals under study. It is an early "exemplar" that was followed faithfully.

B. THE FLEXOR SYSTEM

1. Adductor Group

The muscles comprising this unit are made up of the Musculi gracilis and the adductores. They are generally innervated by the Nervus obturatorius, and the twigs of the Nervus ischiadicus contribute.

MUSCULUS GRACILIS

Didelphis (Plate 1-j: Fig.3)

Origin: From the ventral surface of the symphysis of the pelvis and the cranial end of the horizontal pubis in broad, thin muscle fibers all of which are the same length as the cell bundle.

Insertion: By a flat, narrow tendon on the ventral or anterior border of the tibial crest and the middle side of the leg. The muscle lies on fascia in this area along with the insertions of the Musculi sartorius, semitendinosus, and the semimembranosus.

Tachyglossus (Plate 2-h: Fig. 4)

Origin: Directly opposite its twin, from the whole length of the symphysis pelvis, and most of the outer surface of the marsupial bone (a sesmoid, *i.e.*, Os epipubis, Fig. 16; Plates 10a-e; 10b-f), and some of the horizontal pubis cranially. It also appears tucked in fleshily two-fifths along the caudal posterior ridge of the ischiadic arch. From here a slip runs backwards toward the cloaca.

Insertion: On the posterior surface of the tibia; here it appears as two thin strips that converge on a short, flat, broad tendon along the middle third of the shaft.

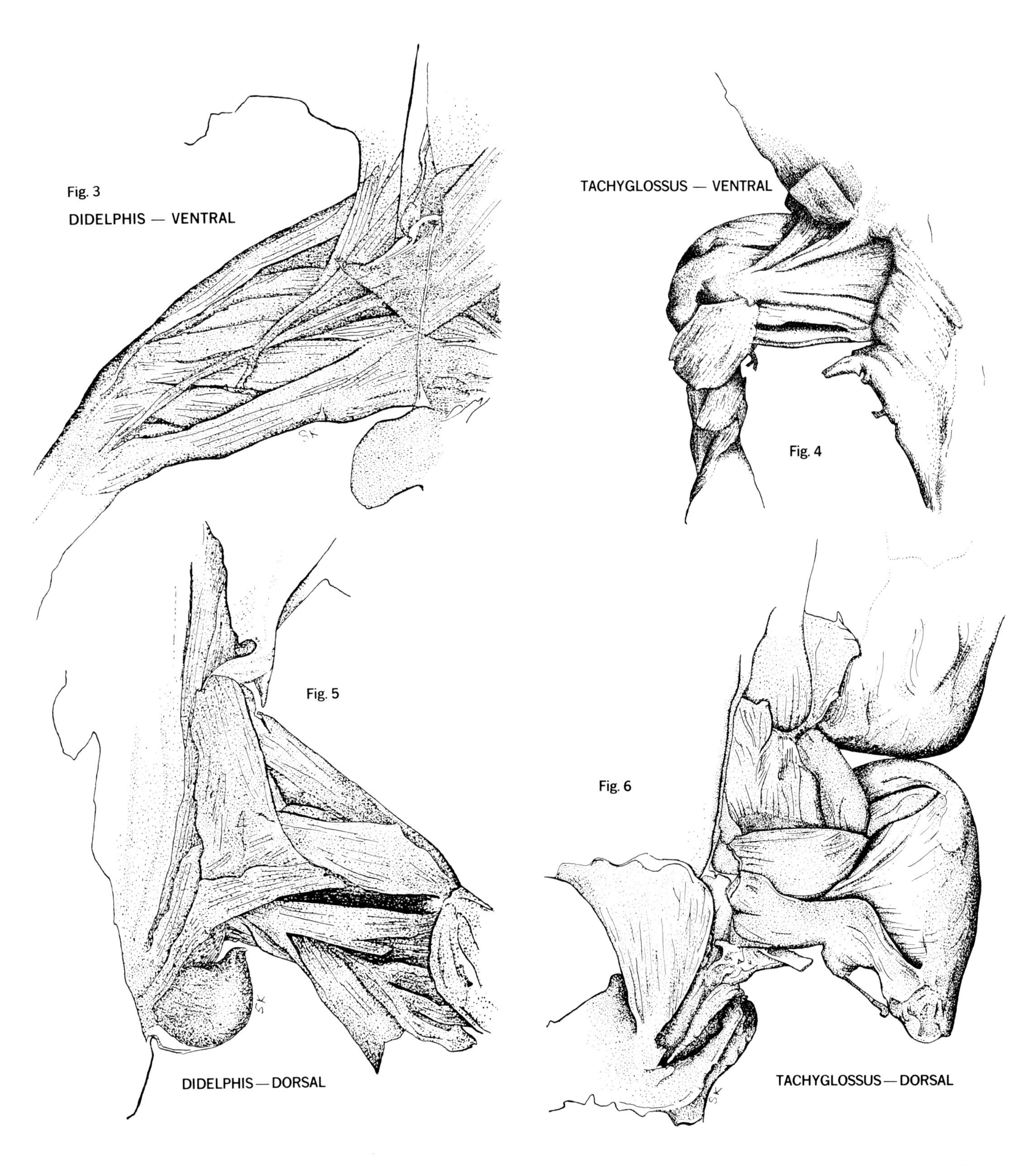
Fig. 3
DIDELPHIS — VENTRAL
TACHYGLOSSUS — VENTRAL
Fig. 4
Fig. 5
Fig. 6
DIDELPHIS — DORSAL
TACHYGLOSSUS — DORSAL

Sphenodon (Plate 5-a; Figs. 8-a; 9)

Origin: Fairly straight on Ligamentum pubioischiadicus and from the pubioischiadicus symphysis, palm-like.

Insertion: By two flat bundles difficult to separate, one overlapping the other, just below the tibial tuberosity and on the tibial crest.

In *Didelphis* this one thin superficial muscle is as usual in true mammals; it is often partly attached in marsupials to the Musculus semitendinosus. In *Tachyglossus* and *Sphenodon* its origin of two masses is usually the same as in the opossum, *Didelphis*; however, in a mature echidna, *Tachyglossus*, its origin does not extend to the epipubic bones. Insertions of all genera vary somewhat. Tendons and fascia lead the insertions to different locations on the tibia—each genera exhibiting specificity of muscle extent. In *Sphenodon* the Musculus gracilis is a fairly thick ventral layer of fibers which appear almost as one sheet, and it is contiguous with the Musculus semimembranosus. In each animal this muscle is fan-shaped with its fibers converging tendinously on the leg.

A branch of the hamstring nerve (Plexus lumbalis), Nervus obturatorius, extends to the insertion of the Musculus gracilis in the opossum, *Didelphis.* Two parts of the same innervate the muscle in the echidna, *Tachyglossus;* while parts of the Nervi obturatorius and ischiadicus send shoots to the reptilian correspondent. All these nerves are part of the lumbar and sacral network.

The action in the opossum, *Didelphis*, is to flex the leg and draw the entire limb toward the body. The Musculus gracilis in the echidna, *Tachyglossus*, and tuatara, *Sphenodon*, has a similar function. Comparatively many more accessory tendons are apparent in tuatara's (*Sphenodon's*) anatomy.

Newman (1877) finds tuatara's Musculus gracilis but it is considered to be the pubioischiotibialis of the generalized reptiles in *Sphenodon* by some other researchers (Byerly, 1925; Romer, 1922, 1923, 1970; Hyman, 1944). In all, nerves run to the sacrum and to the Musculus obturatorius. In the lizard the homolog is usually found originating at the base of the girdle and inserting on the under surface of the tibia. Romer (1922), Gadow (1882), and Gregory and Camp (1918) show in lizards an origin from the pubioischiadic ligament. Gadow (1882) does not present this as a muscle in the alligator. Romer (1923) has a pubioischiotibialis in the crocodile and also *Sphenodon*. Osawa (1898) found it originating on the ischiopubic symphysis in *Sphenodon*; it fitted into the medial side of the proximal end of the tibia. He termed it "Pubio-ischio-femoralis," innervated by the Nervus obturatorius. Gregory and Camp (1918) have the same innervation in tuatara but call it "Pubio-ischio-tibialis." Gadow (1882) considered it one of five sections of the reptilian Musculus flexor tibialis internus. In the turtle (Noble and Noble, 1939) it is called the Musculus gracilis and finds its beginning on the pelvic ligament with a narrow tendon attachment. The insertion is on the inside of the proximal head of the tibia. The Musculus gracilis bends and slightly rotates the leg in most reptiles.

A definite basic pattern of morphology is evident for Musculus gracilis in the three primitive genera at times, irregardless of the function of this muscle in each case.

MUSCULI ADDUCTORES

Didelphis (Plate 1-e,f,g,v: Fig. 3)

Origin: In four parts; one behind the edge of the iliopectineal eminence, which is a division (e) in front of and a little anterior to Musculus pectineus (t); another on the inner margin of the horizontal pubis; one on the symphysis pubis, and

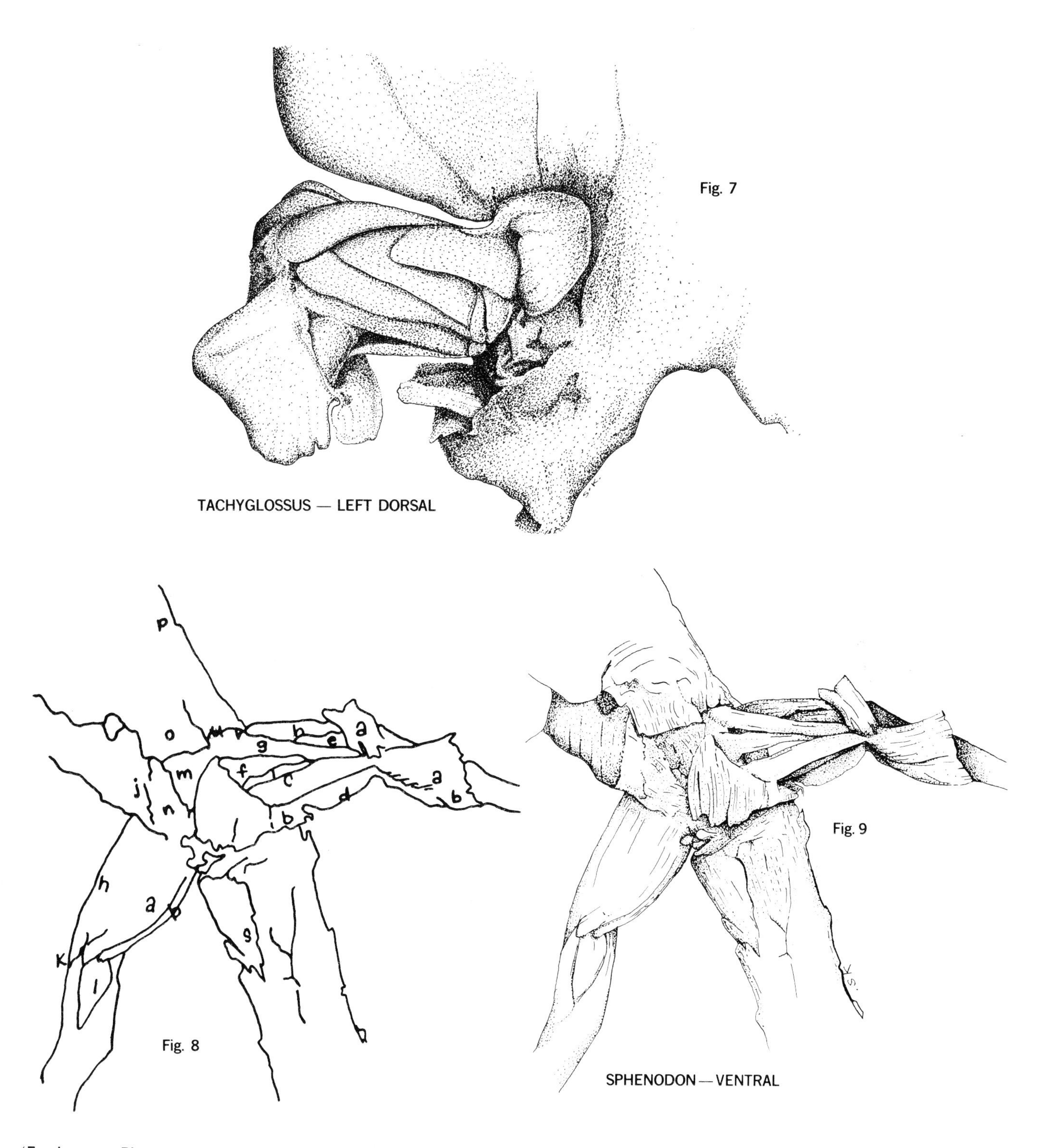

Fig. 7

TACHYGLOSSUS — LEFT DORSAL

Fig. 8

Fig. 9

SPHENODON — VENTRAL

(For key, see Plate 5, page 52)

another from the ischiadic arch as far as the tuberosity-angle of the ischium. The latter encircles the obturator foramen for about 350 degrees.

Insertion: Fleshily into the ventral body of the femur from the lesser trochanter, along the linear aspera as far as the medial condyle and the epicondyle, on the caudal end of the bone.

Tachyglossus (Plate 2-e,g: Fig. 4)

Origin: These separate muscle masses come from the base of the epipubic bone on the horizontal pubis, the symphysis pelvis, and along a fourth of the ischiadic arch toward the ischial tuberosity. They encircle the edge, matching the obturator foramen, and at an angle of 140 degrees.

Insertion: Entirely on the distal third of the median edge of the femur on the medial condyle with a small tendinous connection.

Sphenodon (Plate 5-c,f,g: Figs. 8-c,f,g; 9)

Origin: Three parts; one from the connective tissue attached to the prepubis; another from the pubio-ischiadic ligament along with the Musculus gracilis; a third part arises from the lateral pubic tuberosity.

Insertion: Proximally below the lateral condyle along with the Musculi semimembranosus and semitendinosus. Two divisions attach to the underside of the femur. The other part inserts on the inner side of the tibia.

In the marsupial, *Didelphis*, the four-parted adductores arise as longitudinal fibers from the pubis and ischium and have a long insertion. Because some of the echidnas, *Tachyglossi*, were immature animals, three separate adductores as a group are seen less extensive than in *Didelphis*. Three narrow strips of muscle appear as these muscles in *Sphenodon* originating from the pubio-ischiadic ligament. The reptilian pubiotibialis in rhynchocephalia, *Sphenodon*, seems to be an adductor of the thigh. This special muscle, termed the same in most reptiles, is not supposed to have a homolog in mammals (Romer, 1923). Even if not, it is a true adductor here.

In the opossum, *Didelphis*, the Nervus obturatorius and ischiadicus help form the neuromuscular juncture of the Musculi adductores. The echidnas, as in lizards and tuataras, have the same double innervations. This is not so in the eutheria, the placental mammals.

In the opossum, *Didelphis*, these masses adduct and extend the hip, turn out the thigh. The large size of these adductores and a wide acetabulum permit great freedom of motion of the femur, an arboreal adaptation. In the echidna, *Tachyglossus*, the adductor group extends and draws the leg toward the body. The muscles move or pull toward the median axis in a clockwise circular motion. As they do this, they push backward. As an animal digs with the front feet, the hind legs serve as the support and the curved hind claws move in a circular motion counterclockwise (Jenkins, 1970b). In tuatara, *Sphenodon*, the action is similar to an adductor of the thigh, but is also the flexor of the hip. This latter is particularly true of the Musculus adductor (Plate 5-g).

Byerly (1925) considers this "Pubo-ischitrochantericus" and "Ischio-femoralis" as adductores of the thigh in *Sphenodon*. Gregory and Camp (1918) in this same tuatara, label this muscle part of pubio-ischio-femoralis. Romer (1922) mentions "Ischio-femoralis" as an adductor in generalized reptiles. Romer (1942) has an adductor femoris in *Lacertalia* and *Sphenodon*. He also finds the above mentioned pubio-tibialis in these two reptiles. Hyman (1944) has adductores as hind-limb parts derived from the sources of primitive ventral muscles, but considers these (Musculus semimenbranosus, the Musculus semitendinosus and the Musculus biceps) Musculi flexor tibialis internus and externus, and pubio-

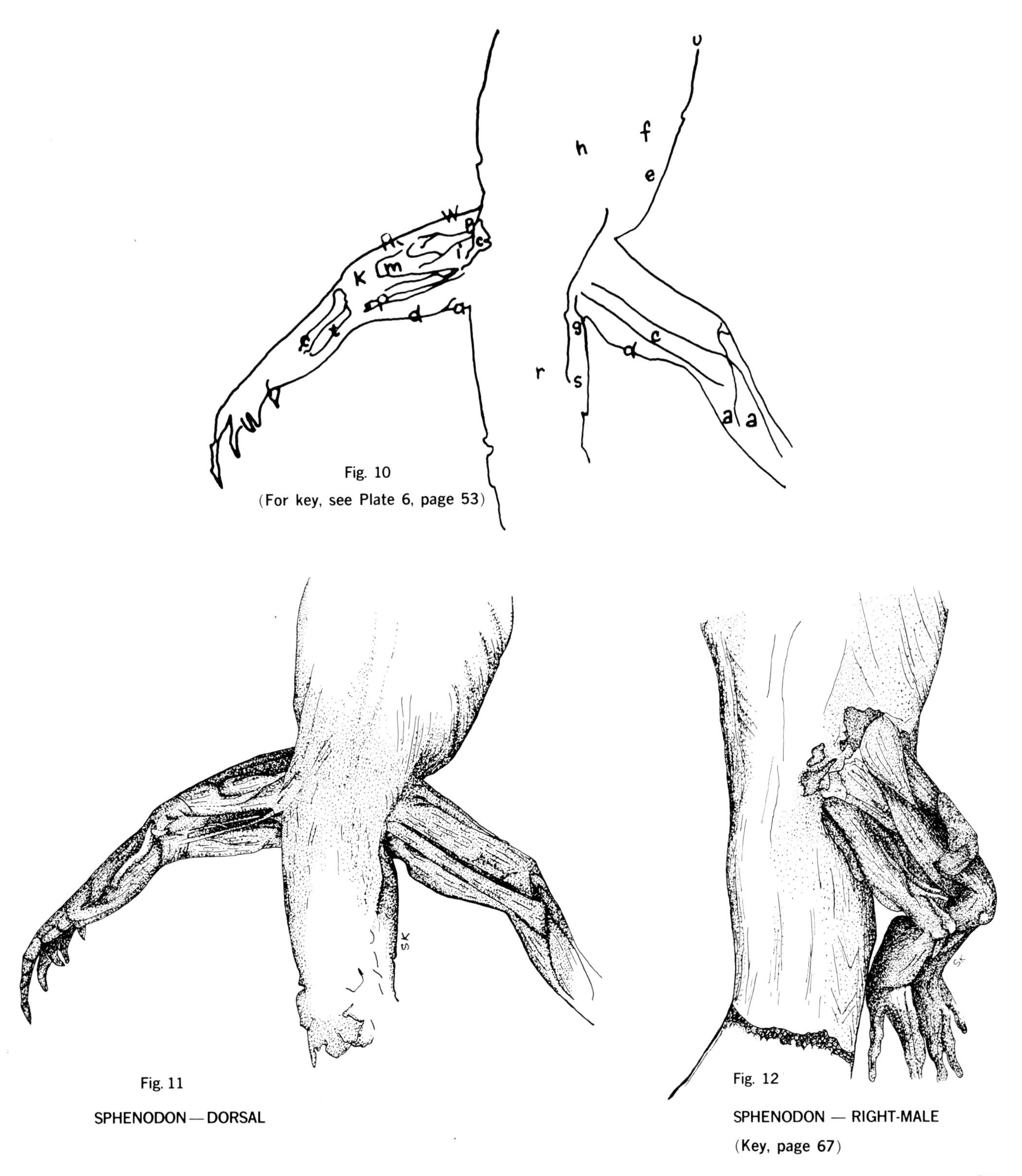

Fig. 10

(For key, see Plate 6, page 53)

Fig. 11

SPHENODON — DORSAL

Fig. 12

SPHENODON — RIGHT-MALE

(Key, page 67)

tibialis. Gadow (1882) and Osawa (1898) call adductors in *Sphenodon*, "Pubi-ischio-tibialis."

Gregory and Camp (1918) think the adductor series probably represent the more superficial segments of the pubi-ischiofemoralis externus of other primitive reptiles. In the alligator, Gadow (1882) calls this the ischiofemoralis, innervated by the obturator nerve. In the lizard, Gadow (1882) and Osawa (1898) homologize the adductor group with the pubio-ischiotibialis. Romer (1970) calls the mass Musculus adductor femoris in the lizard. A paper by Romer (1923) showed that the chief differences between the long flexors of the lizards and crocodiles were that the pubo-tibialis (Musculus adductor in *Sphenodon* in this paper) was missing, and that the ilio-ischiadic ligament was lost; in consequence, the mass Romer called pubo-ischiadic ligament was smaller. So the alteration of the pelvic plate is found in the crocodile changed from the lizard condition. Gadow (1882) did not demonstrate a pubio-ischiotibialis (Musculus gracilis) in the crocodile. Synonyms and homologs again are so often disputed in the early literature. Chiasson (1962) labels a Musculus adductor femoris in the alligator. Its origin is found on the anterior border of the ischium and insertion is on the anterior distal half of the femur. In the turtle there is a triceps femoris adductor (Noble and Noble, 1939; Owens, 1866) which has its inception on the ventral surface of the pubis and ischium and fits into the proximal head of the femur. Its action is extension of the leg. In the turtle, Ashley (1955) also has a Musculus triceps femoris adductor beginning on the pubis, ventrally, and on the ischium. It also runs into the proximal head of the femur.

2. Hamstring Group

The muscles here are composed of the Musculi biceps femoris, semimembranosus, semitendinosus. (In general, they receive their innervation from the Nervus tibialis and the end branch of the sciatic nerve—Nervus ischiadicus. Sometimes, the Plexus lumbalis is involved.)

MUSCULUS BICEPS FEMORIS

Didelphis (Plate 3-c: Fig. 5)

Origin: On a tendon with the Musculus semitendinosus from the tuberosity of the ischium extending from a single long head.

Insertion: On the lateral surface of the leg, around it as far as the crest of the tibia and part of the Ligamentum patellae on muscle connective tissue around the fibula.

Tachyglossus (Plates 2-i; 4-f,v: Figs. 4; 6)

Origin: From the pelvic arch and the tuberosity of the ischium.

Insertion: On the fascia covering the muscles of the leg; the wide band extends from the tibial crest to the area just above the medial malleolus. It wraps around the bone.

Sphenodon (Plate 6-c: Figs. 10-c; 11)

Origin: On the ventral edge and side of the wing of the transverse process of second sacral vertebra.

Insertion: On lateral surface of the proximal edge of the head of the fibula, below the Musculus gastrocnemius, between the bellies where this muscle separates.

The Musculus biceps femoris on the American opossums dissected had one head. Most marsupials show two parts (Barbour, 1963).[12] The Musculus crurococcygeus in one opossum, *Didelphis*, has some insertion with the Musculus biceps femoris. It seems to have no connection whatever with the femur, and does not attach to

[12] Barbour, R. 1963. Aust. J. Zool., 11:585.

PELVIS
SPHENODON

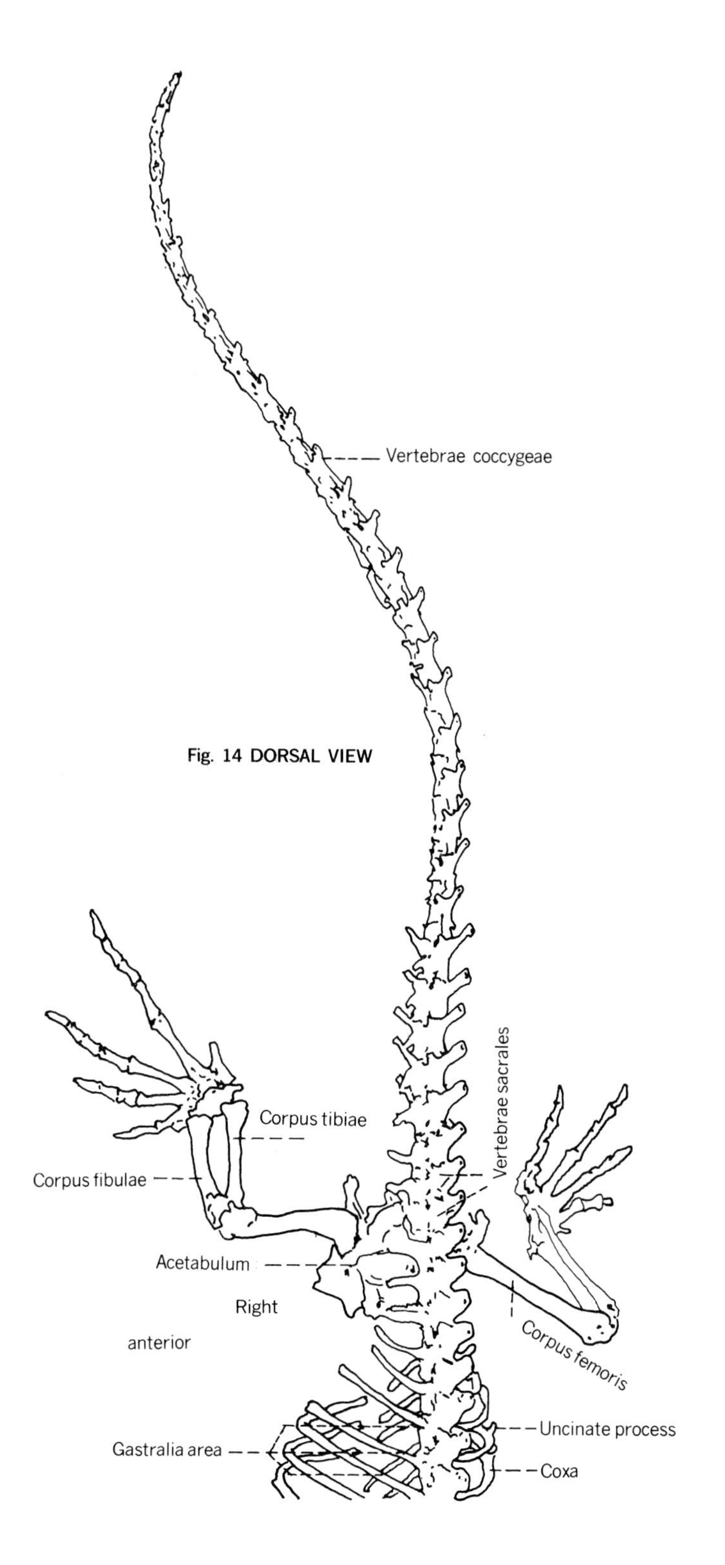

Fig. 14 DORSAL VIEW

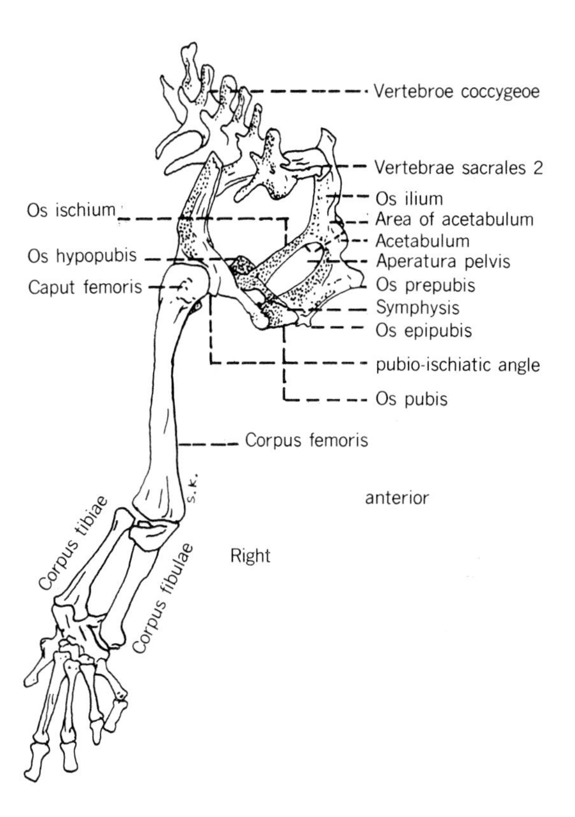

Fig 13 DORSAL DETAIL

PELVIC ARCH OF SPHENODON

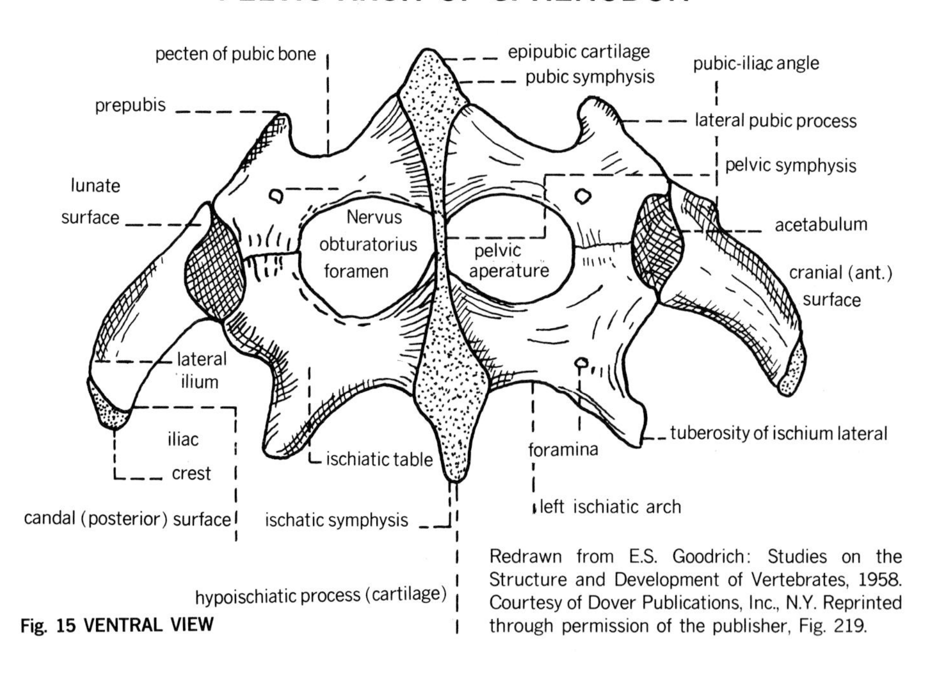

Fig. 15 VENTRAL VIEW

Redrawn from E.S. Goodrich: Studies on the Structure and Development of Vertebrates, 1958. Courtesy of Dover Publications, Inc., N.Y. Reprinted through permission of the publisher, Fig. 219.

the fibular except on fascia. In some situations the biceps has its upper border in close relation with the Musculus femorococcygeus, while its lower edge is attached to the Musculus caudifemoralis (dorsal head of the semitendinosus). In the echidna, *Tachyglossus*, this muscle is fused with the Musculus semitendinosus at its insertion. Part of the Musculus biceps femoris in the same position in tuatara, *Sphenodon*, is inseparable, as it extends to the fibula. In all of the first two genera above, this is a single mass, with fibers blending into broad sheets of connective tissue. In the rhynchocephalian, *Sphenodon*, the muscle is a narrow strip with no connective tissue attachment. In a primitive reptile Osawa (1898) has this muscle arising from the lateral surface of the ilium, above the origin of the Musculus gluteus (ilio-femoralis). This was not true in the *Sphenodon* dissected. Note the arrangement of the originating fibers (Plates 23-3; 24-3).

Romer (1922) called this the Musculus iliofibularis in the general reptilian condition; it was termed the same in the crocodile (Romer, 1923)—with sacral innervation; but in this advanced reptile it had the same relation with Musculus flexor tibialis internus, (Musculus semimembranosus), and it ended in the tibia. In his study of the lizard (1942), Romer found that the homolog of the mammalian biceps was similar, *i.e.*, the Musculi iliofibularis and flexor tibialis internus were muscles in two parts with a sacral innervation. Later it was just the Musculus flexor tibialis internus (Romer, 1970). In modern reptiles, as in the primitive animals, it has no connection with the femur. Generally its source is on the ilio-ischiadic ligment in lizards, which runs to the tuberosity of the ischium (this ligament is lost in the monotreme, *Tachyglossus*, and marsupial, *Didelphis*). It is introduced into the fibula. Osawa (1898) in the primitive *Sphenodon* calls this ilio-fibularis, originating on the lateral surface of the ilium and ending on the proximal lateral surface of the fibula. Here it is innervated by part of the Nervus peroneus (fibularis). Gadow (1882) in the alligator shows two or three heads of the Musculus ilio-fibularis, and two heads of the Musculus flexor tibialis internus in the lizard.

Furbinger (1874) confirmed this in the lizard also. Gregory and Camp (1918) called this the Musculus ilio-fibularis; to them the Musculus flexor tibialis internus was the Musculus semimembranosus. Chiasson (1962) in the alligator shows both a Musculus iliofibularis and a Musculus flexor tibialis internus. The former arises from the crest of the ilium and inserts by a long narrow tendon into the anterior proximal end of the fibula. The Musculus flexor tibialis internus comes from the posterior margin of the ischium and part from the dorsal posterior section of the ilium. It inserts on a tendon which leads to the fascia on the medial surface of the proximal end of the shank. The turtle is another story. Ashley (1955) and Noble and Noble (1939) demonstrate no Musculus biceps femoris, no Musculus iliofibularis, no Musculus flexor tibialis internus. The flexion of the shank is reduced; it has been taken over by other muscles. The shells have limited the movements in *Chelonia*. The changed position of the biceps from the deleted reptilian ilioischiadic ligament is lost and obviously, as is generally known, has to do with the differences in the position of the limbs (Fig. 2).

In summary, in the opossum, *Didelphis*, the Musculus biceps femoris extends the hip, bends the knee and throws out the thigh. In the echidna, *Tachyglossus*, it pulls back the thigh, it extends and bends the leg, and it rotates the limb below the knee. The heel and the claws turn outward away from the body counterclockwise. Tuatara's, (rhynchocephalian's) *Sphenodon's*, action is extension, but this movement is quite limited in this muscle. However, it helps turn the leg on its axis.

In general, the nerves supplying Musculus biceps femoris in these primitive organisms are from the Plexus sacralis. Again for this region, the archetype is maintained, but the functional role is not always the same.

MUSCULUS SEMIMEMBRANOSUS

Didelphis (Plates 1-h; 3-b,r,u: Figs. 3; 5)

Origin: Single head from the caudal border of the Os ischium, 2.4 centimeters along the Ramus ossis ischii, nearly to the medial angle of the Tuber ischiadicum. A Musculus presemimembranosus is noted (Plate 3-t).

Insertion: On a small tendon with Musculus femorococcygeus on the medial side of the Tuberositas tibiae.

Tachyglossus (Plate 2-m: Fig. 4)

Origin: From the entire lateral angle of the Tuber ischiadicum.

Insertion: Near the tibial crest by a small ligament along top side of the anterior mesial border of the tibia.

Sphenodon (Plate 5-d; 6-d: Figs. 8-d; 9; 10-d; 11)

Origin: On connective tissue with Musculus femorococcygeus at tip of the first caudal vertebra. It straddles this muscle, the fibers join with the Musculus semitendinosus on the edge of Tuber ischiadicum about one fifth of the way along the Arca ischiadicum.

Insertion: By a tendon on tibial tuberosity; it wraps over the trochanter just above the insertion of Musculus iliopsoas and Musculus pectineus.

Although it has a single head in the opossum, *Didelphis*, sometimes this muscle seems to be partly split into two, one of which is a Musculus presemimembranosus, supposedly a muscle in the

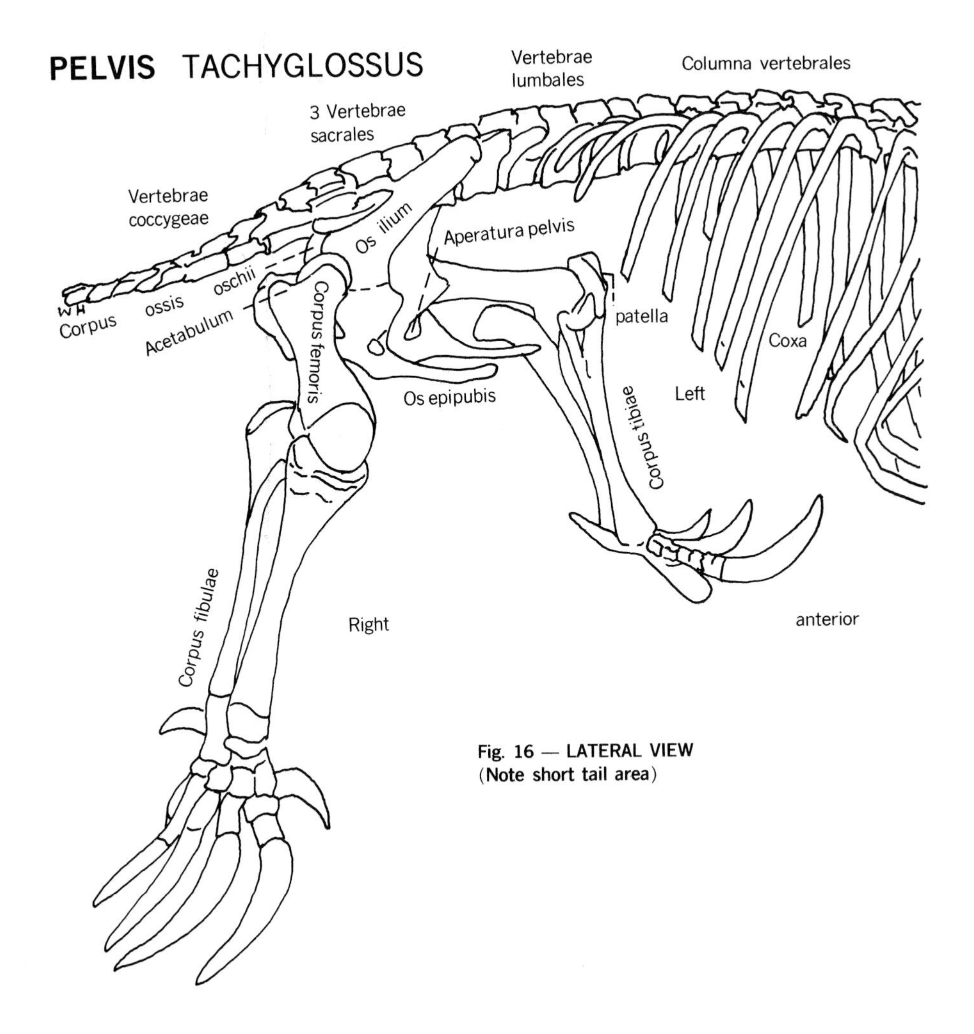

Fig. 16 — LATERAL VIEW
(Note short tail area)

process of fusing with the single-headed Musculus semimembranosus (Plate 3-b,t: Fig. 5). However, in the echidna, *Tachyglossus*, its head appears as a fusion of two or three muscles, very closely associated with the Musculus semitendinosus. The bundle in tuatara, *Sphenodon*, demonstrates origin from fascia on which lie beginnings of other fasiculi. Partial splitting is noted and fibers meld with Musculi semitendinosus and femoro-coccygeus, making it almost impossible to separate one muscle from the other.

A hamstring nerve, Nervus ischiadicus, innervates this bundle in the opossum, *Didelphis*. In the echidna and tuatara, which have some common reptilian characteristics, the same nerve persists but there are small branches from the Plexus lumbalis, two innervations similar to all the reptilian flexors. In his embryonic specimens, Low (1929) found only Nervus ischiadicus at the neuro-muscular junctions. In all three genera the adaptations for function as flexors of the thighs are the same. They extend the hip, bend the knee and turn the shank inward. In burrowing animals like *Tachyglossus* and *Sphenodon*, the Musculus semimembranosus is well developed.

Depending to whom the references belong, the

Musculus semimembranosus of primitive animals above is synonymous and homologous with the ischio-tibialis of reptiles, or the flexor tibialis internus. Gadow (1882) and Gregory and Camp (1918) call it flexor tibialis internus in the alligator. Osawa (1898) and Gadow (1882) in the *Sphenodon* reptile, term the medial portion, "ischiotibialis-posticus," which was equal to the flexor tibialis internus. Romer (1922) considers the long flexors as flexor tibialis internus or ischioflexorius, to be the general reptilian condition. Byerly (1925) calls this muscle "Ischio-tibialis posticus." In the alligator there is no Musculus semimembranosus noted by Chiasson (1962), but a flexor tibialis internus is shown which was mentioned above under the remarks about the Musculus biceps femoris. The turtle has a Musculus semimembranosus (Noble and Noble, 1939; Ashley, 1955). Owens (1866) has a Musculus semimembranosus with two origins, and insertion in the tibia. Romer (1923) explains in his study of the crocodile, that in this most advanced reptile, with muscle parts broken off, the homology of the Musculus semimembranosus is troublesome. He concludes, from an investigation into the Lacertalia (1942) and primitive reptiles, that the flexor tibialis internus with its obturator and sacral innervations has the same structure and relationship. In a final decision (1970), Romer suggests that the under surface of the thigh musculature, the Musculi flexor tibialis internus and externus, are the same as the semimembranosus, semitendinosus, and perhaps the biceps also. Hyman (1944) made the same conclusion.

Except for *Sphenodon's* myology here being more extensively draped on fascia, the same fasiculi in *Didelphis* and *Tachyglossus* show basic design. This is an example again of reptilian appendicular muscles which have developed into powerful muscles in higher tetrapods.

MUSCULUS SEMITENDINOSUS

Didelphis (Plates 1-r; 3-a: Figs. 3; 5)

Origin: Caudal, from the beginning and close to the Musculus biceps femoris, on the Tuber ischiadicum; a dorsal and ventral division or separation in some opossums.

Insertion: On the crus in front of the Corpus tibiae 3 cm below the knee on a short tendon. There can be three insertions together, parts of different muscle bundles.

Tachyglossus (Plate 2-j,i,n: Fig. 4)

Origin: With the Musculus semimembranosus and behind it dorsally from border of the Musculi obturatorius, externus and internus, and from the Tuber ischiadicum.

Insertion: On the dorsal surface of the tibia, along its mesial edge, below the Musculus semimembranosus.

Sphenodon (Plate 5-b; 6-q: Figs. 8-b; 9; 10-q; 11)

Origin: Outer surface of the Tuber ischiadicum, and Arca ischiadicum on caudal connective tissue, and tendon of Musculus femorococcygeus.

Insertion: On the medial lateral proximal end of tibia along with the insertion of Musculus semimembranosus–Musculi gastrocnemius.

In the opossum the dorsal head of this muscle called by Romer and others, the Musculus crurococcygeus, fuses with the ventral part of this Musculus semitendinosus and/or with the biceps. Romer (1922) shows this fusion inserting internally on the tibia. This muscle in two specimens of our opossums is along the Musculus biceps; it crosses the Musculus semimembranosus diagonally and goes into the lower part of the leg. It is constricted with the Musculus crurococcygeus (its dorsal head) about half way along its length; this may be an example of the fusion of unrelated muscles in evolution. Coues (1871)

PELVIS TACHYGLOSSUS

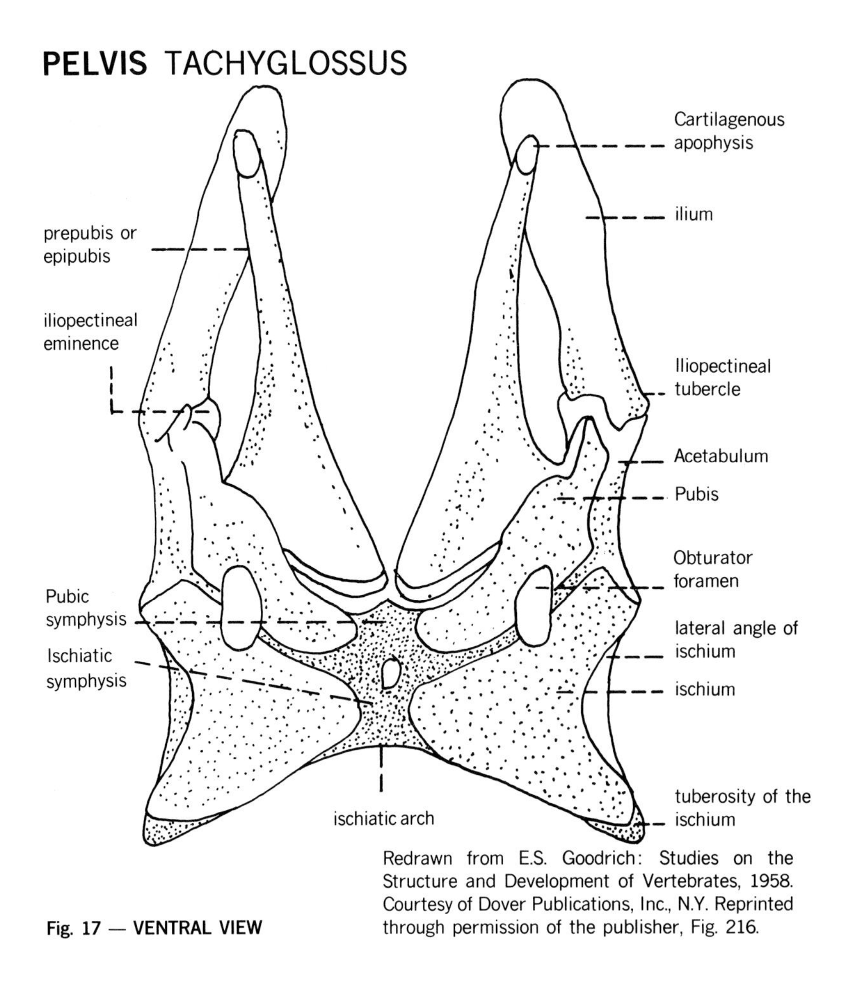

Fig. 17 — VENTRAL VIEW

Redrawn from E.S. Goodrich: Studies on the Structure and Development of Vertebrates, 1958. Courtesy of Dover Publications, Inc., N.Y. Reprinted through permission of the publisher, Fig. 216.

finds a reptilian Musculus intertibialis in the platypus monotreme, *Ornithorhynchus*, which has fibers that inserted into the semitendinosus—this muscle is closely associated with the semimembranosus as also in the monotreme, echidna. In the dissected specimens of tuatara no separation of the semitendinosus into separate bundles was noted. The origin persists from the Os ischii and here it is difficult to separate out single masses. The groups form a flat, thick, folded enlargement, which is made of many muscles; the large broad piece curves around the femur to the dorsal side. In all three genera, this muscle with Musculus semimembranosus flexes the leg.

The nerve leading into these muscles in *Didelphis* is branches of Nervus tibialis (Nervus ischiadicus). Nervus isischiadicus and Nervus obturatorius are the innervations noted in *Tachyglossus* and *Sphenodon*.

Osawa (1898) called this muscle Musculus ischiotibialis porticus, a lateral portion, in the primitive reptile. To Gadow (1882) in the same animal it was the flexor tibialis externus. In lizards and *Sphenodon*, it was given the same name by Romer (1942, 1970) and it had sacral

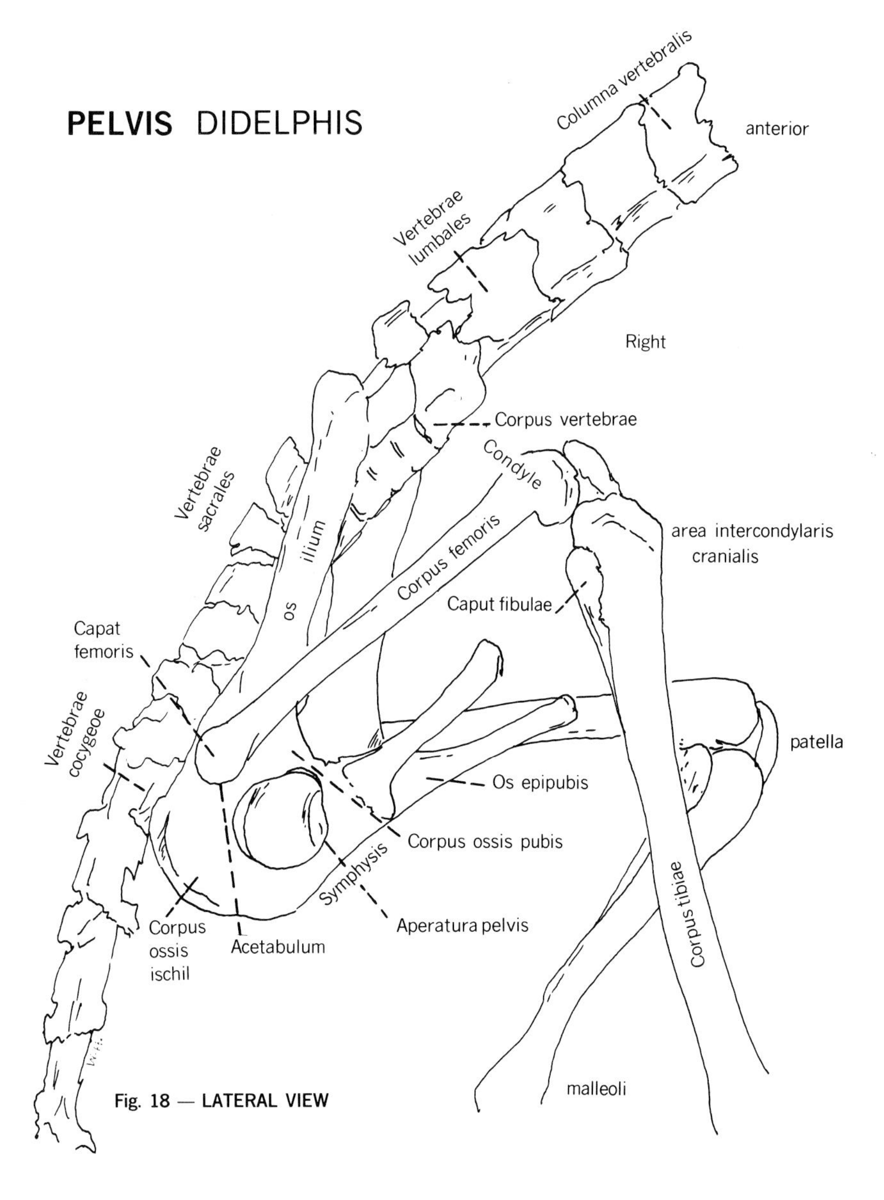

Fig. 18 — LATERAL VIEW

innervation; this is verified in the Crocodilia (Romer, 1923; Gadow, 1882). Gregory and Camp (1918) has the semitendinosus equal to "caudo-flex" which they say is like the flexor tibialis externus in primitive reptiles. Hyman (1944), as stated above, classifies these hind limb primitive muscles, *i.e.*, Musculi pubo-tibialis, flexor tibialis internus, and flexor tibialis externus as Musculi semimembranosus, semitendinosus, and biceps. Chiasson (1962) mentioned a flexor tibialis externus. In the turtle no notice of this muscle is found in our references except by Owens (1866). His Musculus semitendinosus has three origins, one from the back part of the higher end of the ilium; a second from the tuber ischiadicum; the third from the Os ischii and Facies symphyialis. They insert on the tibial tuberosity and within the Musculus gastrocnemius.

IV DISCUSSION

What is known about reptile, monotreme, and marsupial gross morphology, which molds the construction of the bones of these primitive animals, is scattered in references to monotremes and marsupials throughout the literature for over a century and a half. It is found in general papers, specific monographs, dissertations and textbooks from around 1860, and some after the turn of the century; Mivart, 1866—monotremes, marsupials, reptiles; Coues, 1872—marsupials, 1871—monotremes; Westling, 1889—monotremes; Gregory and Camp, 1918—reptiles, monotremes; Osgood, 1921—marsupials; Pearson, 1926—monotremes; Romer, 1922, 1923—reptiles, marsupials, monotremes; Langworthy, 1932—marsupial; Appleton, 1928—tetrapods; Owen, 1866—tetrapods; Low, 1929—monotremes; Elftman, 1929—marsupial; Howell, 1936, 1938—tetrapods; Enders, 1937—marsupial; Barbour, 1963—marsupials; Kerr, 1955—marsupial; Block, 1964—marsupials; Lewis, 1963—monotremes; Griffiths, 1968—monotreme; and others. Besides the reptilian references cited above, work on the rhynchocephalia has been found in Gunther, 1867; Newman, 1877; Osawa, 1898; Gadow, 1882; Perrin, 1895; Romer, A., 1942, 1964, 1966; Romer, F., 1898; Furbinger, 1874; Haines, 1934, 1935, 1938; Schauinsland, 1900; Byerly, 1925; and others. This list includes the work relating to the pelvic and thigh musculature. The information includes some study of the literature about *Ornithorhynchus*, known commonly as the platypus, but the muscle analysis for the most part comes from *Tachyglossus*, the common or uncommon echidna! The literature about *Trichosaurus* (Australia), *Didelphis* (North America), and *Caenolestes* (South America) was used as information about comparable marsupials. The *Sphenodon* investigation has the most detailed space.

Here can be pointed out the reptilian system of superficial muscles related to the dissections of the posterior girdle region presented in the echidna and opossums. In general, within any particular group of organisms, the more basic characteristics will be the ones that have evolved more slowly and tend to be more conservative, *i.e.*, moderate; those which lead to or are directed toward rapid evolution are considered to be labile, *i.e.*, unstable.

It is important to trace the course of characteristic muscles which are preservative and the ones that are more likely to change. When a special muscle structure occurs in fundamentally the same form in a great variety of only distantly related organisms (reptiles, monotremes, marsupials), the characters of that muscle structure are conservative and so provide more information about the primitive condition than would characters of some muscle structure that varies greatly from species to species. It is a matter of usefulness of muscles under its own living conditions. Altered structure is unique to a narrow way of life as exemplified in *Tachyglossus*. There are strong fossorial adaptations in the monotreme, mild arboreal ones in *Didelphis*. But the latter is the more generalized animal, meaning broadly adjusted to greater variety of diverse kinds of environment or ways of life. In both those organisms and in *Sphenodon*, through the ages, isolation, lack of competition, small size, have been among the many factors that have helped to maintain the original morphology, while individually, they have developed special unique anatomy. Only in *Didelphis* is the life span short.

What are the muscles that have remained primitive? It is commonly known that mammals have reptilian ancestry. In the late Cretaceous (Mesozoic) there was the evolution of mammals wide-spread over the earth, with the pouched mammals remaining the more primitive through the ages. What are the muscles of the American opossum that reflect this conservatism? What can be the slowly evolved pattern up to the first fossil monotremes known solely from the early Cenozic era in Australia—outside Australia in the Mesozoic, a petrosal bone (Kermack, 1967)?

In the opossum, I consider the preservative muscles to be iliopsoas and pectineus; and the glutei as a group have their enlargement due to the arboreal adaptation. The sartorius, the rectus femoris and the vasti are surely conservative in *Didelphis*; they have become flexors in addition. The patterns are the same in *Sphenodon*. They have maintained their extensor function since they performed this like purpose in the reptiles. In the echidna, the two separate iliacus and psoas reflect the more modern reptilian condition since, in the primitive *Sphenodon*, they are an undivided mass. In the echidna, part of the pectineus demonstrates a shift in origin; it remains dorsal but it splits. The glutei in the monotreme show the plurality and enlargement over the reptilian condition due to the "use" theory, where survival modifies the genetic material. Here it relates again to the unique anatomy "selected for," with a small bit of specialization, the digging habit. Primitive in form, its plasticity is adapted by its environment. Musculus sartorius in all genera changes little. As in the opossum and tuatara, the Quadriceps Femoris Group have not been subject to drastic transition. The Extensors are plainly primitive.

The Adductor mass shows more advanced reptilian relationship in the echidna than in the opossum for the insertions are more like those in the modern reptiles than like those in tuatara. The Hamstrings have fewer deviations from *Sphenodon* in both *Didelphis* and *Tachyglossus*. The turtle and crocodile of today show more advancement in reduction in this group, with the splitting and fragmentation, and change toward regression.

Finally, since the insertion of nerves in muscles always is commonly understood to have close connection with the action of a muscle, and since the muscle is the end organ of a motor nerve (the neuro-muscular junction), original and typical relation can be claimed by a muscle nerve supply (Bock, 1969; Romer, 1970).

It is known that in the brain or spinal cord, the source from which the nerve fibers come—destined for a certain muscle—must be the same as the embryonic and primitive origin. Generally it is true that axons and cell bodies adopt different nerve strands to get to certain muscles. The change seen in Figs. 3; 5 is due more to muscle migration in the opossum and this is not so in the echidna (Figs. 4; 6). Thus any change from reptilian nerve supply is an alteration in the relative position to muscles. Thus nerves can be added to, or deleted from, or become concentrated toward another muscle mass—this last especially if a muscle has been lost. They can deviate, encroach, transfer, be displaced, give out twigs, have single or double innervations in a muscle (Musculi glutei in the opossum is a single twig; in Tuatara and the echidna, generally double nerves attach to a bundle).

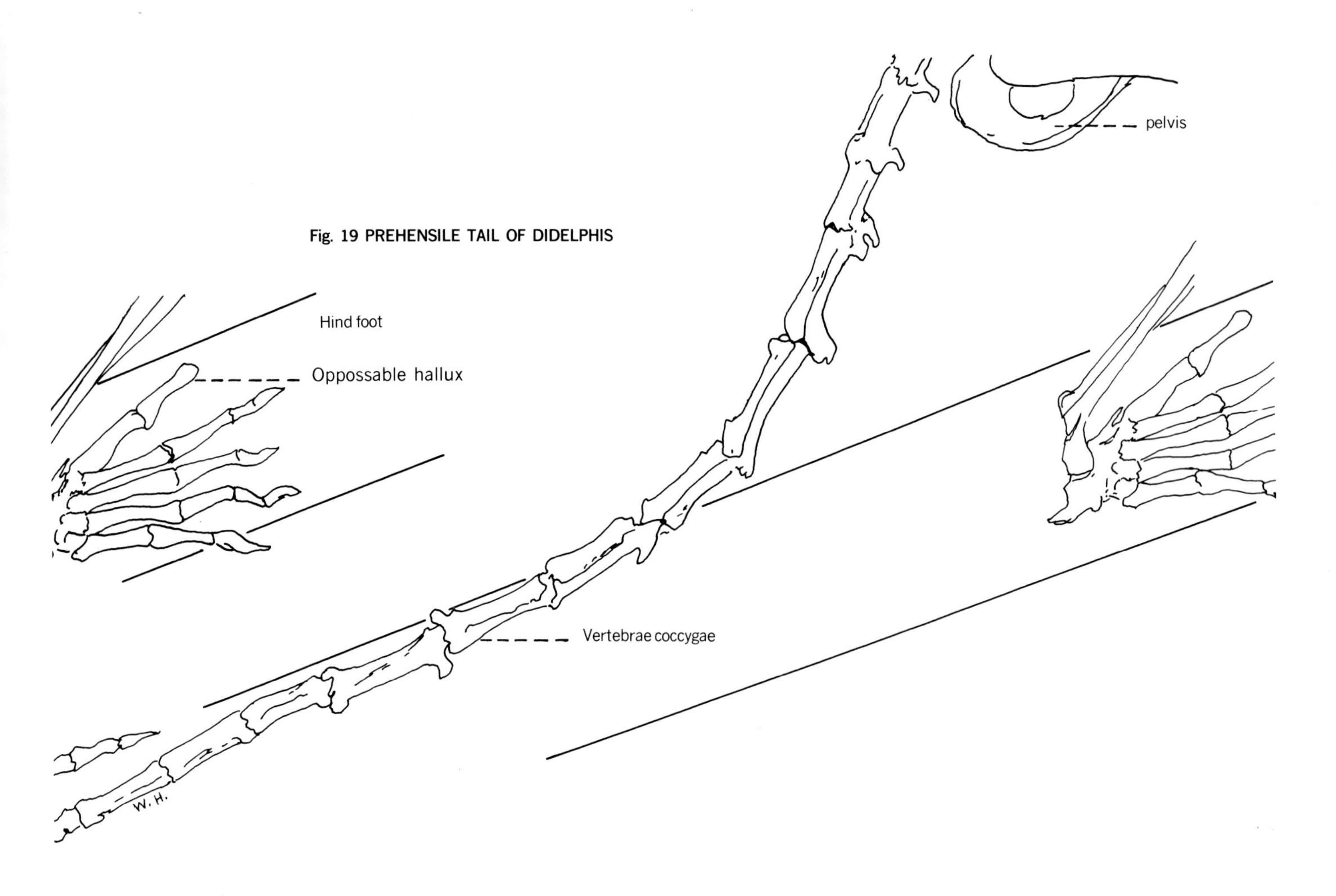

Fig. 19 PREHENSILE TAIL OF DIDELPHIS

Much material has been presented here that is described and asserted by other investigators; much is claimed as positive proof of myological homology. It is hoped that the primitive form and structure that appear universally in *Sphenodon*, *Didelphis*, and *Tachyglossus* will be widely accepted as good reason to use the mammalian nomenclature for muscle architecture in conservative amniotes. The specializations of these animals studied will still be considered as adaptations of form following function. There is thus sufficient fit between materials and the animals' environment. The uniqueness can be accounted for and pointed out in the fossorial uses in the echidna and tuatara of muscles accommodating performance, along with the prehensile tail and the hindlimb opposable hallux (big toe) in the body of *Didelphis*.

What is normal, true, correct, or typical is difficult to point out. Where there is great variation in nerve supply, and where nerve trunks are variable in relation to surrounding muscle structure, evolution has taken place—not so in these conservative amniotes, since preservation in these genera shows primitiveness, and is representative of the reptilian ancestry.

V SUMMARY

The study presented here is of the superficial pelvic muscles of fauna commonly known as living fossils. In these primitive amniotes, the main attention on muscles has been on their place in the developed organism; a study of anatomical units in mechanical movement where comparisons are made between the structures of the pelvis and myology. The muscles that shape the primitive animals—where primitive means more like the archaic lines and where there are many muscles which have not diverged from the basic plan laid down centuries ago—show, naturally, some specializations adapted to the mode of life and environment of each genus.

In the echidna massive posterior muscles are related to digging. The skin is enormously thick, and many papers have been written on its layering and attachment to the superficial muscles of monotremes (Romer, F., 1898; and others). In digging mammals, this thick integument with the girdle serves as a fulcrum, gives leverage, and acts as a brace against which limbs move. Posteriorly, the opossum demonstrates coarse and long muscle bundles modified in the caudal region where the extensive prehensile tail is exceedingly active (Fig. 19). Conversely, the echidna displays a much abbreviated terminal appendage (Plates 4; 4a; 10), with muscle bundles foreshortened but of great depth. In tuatara, the long "metameric-like" muscles betray the primitive condition of a tail whose fasicles work with the legs in propulsion. The comparative myology of the spiny-anteater and the opossum can affirm again the fact that each genus has a tendency to oppose change, and has not been in the process of radiation from a stem form (cotylosauria), even though the primitive mammals arose from the reptilian line. The common American Marsupial and the now protected monotremes have maintained, for their own, certain special places along with the rhynchocephalian; all have remained practically static. The similar superficial muscles seen in the reptiles now existing (especially the lizard, since it is nearer the primitive condition) are not precisely those superficial muscles found in the crocodile and the turtle of today. Despite this, the muscle groups in the pelvic region of these latter orders are definitely tied to the reptiles and therapsids and demonstrate common ancestry. The order, rhynchocephalia, by its muscle and nerve patterns, definitely shows the basic plan of the marsupials and monotremes. These units of common organization are inherited from a shared ancestor of the animal group, and therefore they all have ancient ancestry.

The present investigation also shows, from the attachment of muscles, the position of muscles, and secondarily, from the nerve supply, that muscles in conservative amniotes, in general, retain their relative location with considerable uniformity. The affiliation of fascia, tendons, and ligaments display precise tracks of migration where there is (a), the spread of muscle tissue, increasing in numbers of bundles, attachments, and in size, *i.e.*, Musculi glutei and gracilis in the echidna and the opossum; (b), the resistless move-

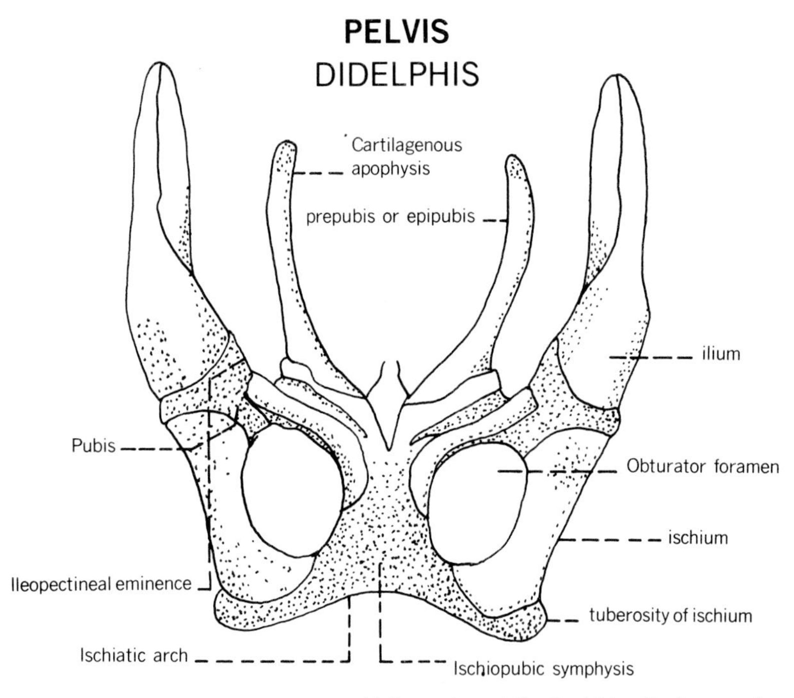

Fig. 20 — VENTRAL

Redrawn from E.S. Goodrich: Studies on the Structure and Development of Vertebrates, 1958. Courtesy of Dover Publications, Inc., N.Y. Reprinted through permission of the publisher. Fig. 216.

ment of a muscle, perhaps turned aside, which is forced to change its position from its derived myotome attachment, due to another muscle wandering in its way, *i.e.*, caudal muscles; (c), the shortening of a muscle, *i.e.*, as Musculus rectus femoris in *Didelphis*; (d), a splitting from the primitive condition, *i.e.*, the Musculus pectineus as a part of the big Musculus iliopsoas-pectineus sheet overlaying the pubi-iliac angle in *Sphenodon* (Plate 5: Fig. 8) fragmenting, a part moving in *Tachyglossus* and *Didelphis*, and a part being retained (Plates 1; 2: Figs. 3; 4).

This study of the controversial question of neuromuscular juncture as criteria of homology leads to some confusion. There is confirmation of the idea (*i.e.*, Romer, 1923), but disputation also (*i.e.*, Haines, 1934; 1938[13]). The latter points up the view that the nerve supply of diverse muscles employs the nerve nearest each bundle in embryonic development, *i.e.*, Musculi adductores and gracilis nerve supplies belonging entirely now to the Plexus lumbalis in *Didelphis*; in *Tachyglossus* and *Sphenodon* the innervations are supplied in the reptilian fashion (double innervation is derived from primitive segmented musculation! *Tachyglossus* and *Sphenodon* Nerve Patterns, Page 17). This does not alter the basic general plan, for it also demonstrates that there can be semblance of the anatomy of nerves in relation to muscles in all three genera.

[13]Haines, R. 1938. Some muscular changes in the tail and thigh of reptiles and mammals. J. Morph., 58:355.

GASTRALIA AND RIBS IN SPHENODON

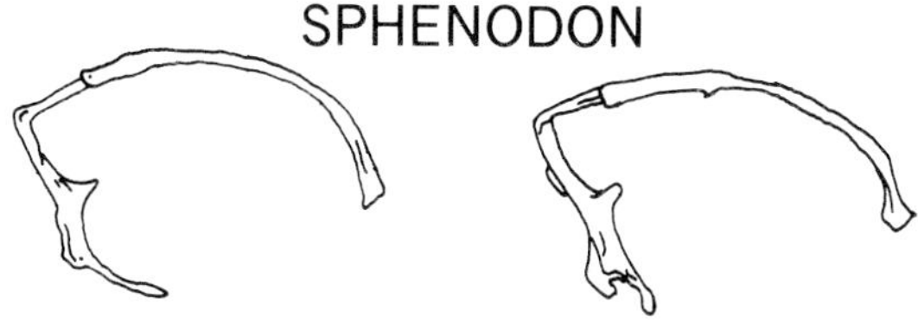

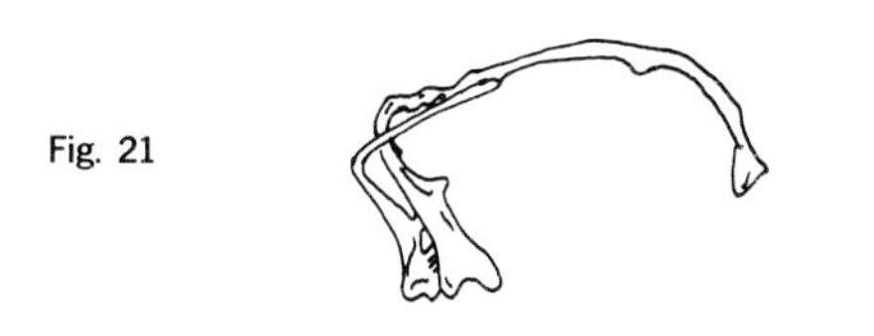

Fig. 21

The resemblance of structural pattern is very close because the accurate and detailed description of insertions and origins are, for all practical purposes, unchanged, and the presence of a nerve with certain relationships to adjacent muscle and bone is a very conservative feature.

It can be concluded that the environment has modified the function and structure of muscles. The changing conditions, the inherited recessive potentiality for conversion, and, or, poor development, really show no disadvantage to the three genera. The environment and the "happy-chance" need have molded the few changing muscle formations in *Sphenodon*, *Tachyglossus*, and *Didelphis*, and consequently, the muscle design for the few specializations.

In summary, the nerve patterns (Page 17) show the incidental changes in the routes of the nerve bundles, especially where there have been muscle cleavage, fractioning, deletion and duplication or replication. They converge, interchange, and blend together slightly. Generally speaking they reflect embryonic and phylogenetic derivation, from my dissections, although this is disputed by some (e.g., Kent, 1965). Innervations are not so diverse in all primitive organisms, because they are in accordance with a definite structural plan. Romer (1922) believes there is definite uniformity in relation of nerve, muscle, and bone.[14] The myological specialization in these conservative amniotes is slight and is due to permutations of attachments, to the transformation of the internal structure of bone where stress has taken over, and to the transmutation of the bone itself which has affected the function of converted muscles. In researching motion, the overall action of pelvic muscles in *Sphenodon, Tachyglossus* and *Didelphis*, in general, is not found changed very much in the old literature, but see Jenkins, (1970b).

Howell (1938)[15] reviews the evolution of development of muscles and bones from *Pisces*, where the ilium has a vertical bar, to where the ilium is slightly oblique in the *Sphenodon* and horizontal to the body in *Tachyglossus* and *Didelphis*. This is interesting to restate because the facts help the understanding that ventral prepelvic musculature, through evolution, became associated with the pubis; the ventral postpelvic bundles closely related themselves with the ischium; the dorsal prepelvic and postpelvic masses of muscle are found linked with the ilium. All the primitive amniote studies follow this course. These conditions are found in all the early fossils of the monotremes, marsupials, and the original rhynchocephalians.

So we see a fundamental resemblance which has not been blotted out, and there can exist

[14]Romer, A. 1922. The locomotion apparatus of certain primitive and mammal-like reptiles. Bull. Amer. Mus. Nat. Hist., 46:523.

[15]Howell, A. 1938. The phylogenetic arrangement of the muscular system. Anat. Rec., 66:311.

homology because there has been presented, in the dissections, anatomic evidence for degrees of relationships and genetic parallel among conservative organisms. However, trends do not continue; how to explain the "stop" in this rare reptile, this protected echidna, and the lone American marsupial in North America, is still unknown to some of the most famous evolutionists (Simpson, 1969, Personal Communication).

The basic pattern of tuatara's myology seems to me to be universally homologous in general, in all three primitive genera, irrespective of similar function, yet this is often the case, and especially in certain situations. Newman (1877), nearly one hundred years ago, must have known human anatomy very well for he applied this same nomenclature to *Sphenodon's* hind limb muscles. Not always does he give to this tuatara the standard name for a muscle as presented in this work, but among those in the pelvic area studied by him are the following: Musculi gluteus, pectineus, sartorius, iliacus, psoas, gracilis, "triceps" (quadriceps) femoris, "femorocaudal" (caudofemoralis), biceps femoris, semitendinosus, semimembranosus, abdominis rectus, gemelli, pyriformis, others—and fascia, *i.e.*, fascia lata, etc.

Perhaps my research on the three genera may help show the way to a complete revision and standardization of myological nomenclature.

GENERAL INFORMATION

In the monotremes and marsupials, the epipubic bones with cartilagenous tips extend anteriorly from the iliopectineal line of the innominate bone. In reptiles, some have thought that the gastralia, which are the ventral abdominal ribs seen in alligators, crocodiles, and tuatara, or perhaps the prepubis of the Crocodillia, or the epipubis of the Chelonia, became calcified in these lines leading from the therapsid mammals. The epipubic bones support weight, and in the heavy extinct dinosaurs, the pelvic area had these bones. The strange reptilian orientation of the pelvis also helped to carry weight, *i.e.*, the pubic processes were directed forward, and the ischium and pubis were parellel. The ventral pelvic plate of *Sphenodon* specifically, bears prepubic and epipubic processes (Plate 8: Figs. 13-15). The short iliums bend just slightly backward.

Theories on the evolution of these Prototheria and Metatheria are varied (Reed, 1960; Simpson, 1959, 1960, 1964; van Valen, 1960; Romer, 1966, 1970; Walker, 1965). The fossils of the monotremes are known only from the Pleistocene Epoch in Australia, while in North America the common opossum is evident from the Cretaceous Period. The general feeling appears to be that the Therapsida (small mammal-like reptiles) were the forebears from which the monotremes are evolutionary lines, separate from all the mammals (Fig.2). The marsupials are considered off-shoots of the Therapsida but arose from primitive Metatheria. The living marsupials took one path (Australian wombat, kangaroo, koala; American opossum; the opossum rats, *Caelonestes*, of South America, etc.) while the Eutherian line led to the placental mammals of today. It is found that the marsupials have diverged the less from the ancestral stock.

Tuataras are found in the early Triassic sediments. The two-arched reptilian skull was known in the early Mesozoic, and *Sphenodon* retained this type of skull, which makes it a separate order among the reptiles, the Rhynchocephalia. The true primitive tetrapods were the early Cotylosauria; it is believed the reptilian line led from this fossil form. So it can be said that along with this common ancestor of the rhynchocephalians, the monotremes and marsupials also diverged. The extinct Pelycosaurs of the early Permian Period, according to the literature (?), are a later offshoot; from them the therapsids, over 190 million years ago, gave rise to monotremes, marsupials, and placental mammals. The fossil history is a very controversial one.

This is not an evolutionary study but one in anatomy. In the following pages are presented the pelvic muscle dissections of the tuatara, *Sphenodon.*

VI. BIBLIOGRAPHY

Appleton, A. 1928. The muscles and nerves of the postaxial region of tetrapod thigh. J. Anat., 62:364-438.

Ashley, L. 1955. Laboratory Anatomy of the Turtle. W. Brown & Co., Iowa.

Baird, D., and W. Take. 1959. Triassic reptiles from Nova Scotia. Bull. Geol. Soc. of America, 76:1565-1566.

Bakkar, R. 1968. Advanced mammal-like reptiles. Discovery, 3:11-22.

Barbour, R. 1963. Musculature and limb plexuses of *Trichosurus vulpecule*. Aust. J. of Zoology, 11:486-610.

Bick, Y., and W. Jackson. 1967. Karyotype of the monotremes *Ornithorhynchus anaticus* (Platypus) and *Tachyglossus aculeatus* (Echidna). Nature, 214:600-601.

Bick, Y., and W. Jackson. 1967a. DNA content of monotremes. Nature, 215:192-193.

Block, M. 1964. The blood forming tissues and blood of the newborn opossum. Ergebnisse der Anatomie und Entwicklungschichte, Band 37, Spring-Ver., Berlin.

Bock, W. 1969. Discussion: the concept of homology. Ann. N.Y. Acad. Sci., 167(1):71-73.

Bogert, C. 1958. The tuatura: why is it a lone survivor? Scientific Monthly, 76:163-170.

Bolk, L. 1894. Beziehungen zwischen Skelett, Muskulatur und Nerven der Extremitaten. Morph. Jahrb., 21:241-277.

Bourne, G. (ed.). 1960. Structure and Function of Muscle. Vol. 1. Academic Press, New York.

Brink, A. 1956. Speculations on some advanced characteristics in higher mammal-like reptiles. Paleontologia Africana, 4:77-96.

Byerly, T. 1925. The myology of *Sphenodon punctatus*. Univ. of Iowa Studies. Pub. by the University, Iowa City.

Carroll, R. 1964. The earliest reptiles. J. Linn. Soc. (Zool.), 45:61-83.

Carroll, R. 1966. Microsaurs from Westphalian B of Joggins, Nova Scotia. Proc. Linn. Soc. Lond., 177:63-97.

Carter, G. 1957. The monotremes and the evolution of mammalian organization. Proc. Zool. Calcutta, Mookerjee Memor., Vol. 1:195-206.

Chiasson, R. 1962. Laboratory Anatomy of the Alligator. W. Brown & Co., Iowa.

Colbert, E. 1955. Evolution of the Vertebrates. John Wiley and Sons, Inc., New York.

Conrad, G. 1940. By boat to the age of reptiles. Natural History, 45:224-231.

Coues, E. 1872. On the osteology and myology of *Didelphys virginiana*. Mem. Boston Soc. Nat. Hist., 2(1):41-149.

Coues, E. 1871. On the myology of *Ornithorynchus*. Proc. Essex Inst., 6:127-173.

Crompton, A. 1964. A preliminary description of a new mammal from the upper Triassic of So. Africa. Proc. Zool. Soc. London, 142:441-452.

Cunningham, D. 1881. The relation of nerve-supply to muscle homology. J. of Anat., London, 16:1-9.

Darlington, P. 1965. Biogeography of the Southern End of the World. Harvard Univ. Press,

Cambridge.
Darlington, P. 1966. Zoogeography. Mus. of Comp. Zoology, Harvard Univ., Cambridge.
Davis, D. 1963. Principles of Mammalogy. Rheinhold Pub. Co., New York.
Diener, E., and E. Ealey. 1965. Immune system in a monotreme. Nature, 208:950-953.
Diener, E., and E. Ealey. 1966. Phylogenetic development of the immune system with special reference to primitive mammals, echidna and platypus. Presented XIth International Congress of Haematology, Sydney, 1966.

Elftman, H. 1929. Adaptations of the marsupial pelvis. Bull. Amer. Mus. of Nat. Hist., 53:189-272.
Ellsworth, A. 1965. Muscle dissections of the American opossum. (unpublished).
Ellsworth, A. 1966. *Didelphis marsupialis virginiana* Dissections. M.S. Thesis Univ. of Conn., Storrs.
Enders, R. 1937. Panniculus carnosus and formation of the pouch in *Didelphis*. J. Morph., 61:1-26.

Fewkes, J. 1877. Contributions to myology of *Tachyglossus hystrix. Echidna hystrix.* Bulletin Essex Institute, Salem, 9:111-137.
Flower, W. 1885. An Introduction to Osteology of Mammalia. (Original pub. by MacMillan and Co.). New 3rd edition rev. with H. Gadow, 1966. A. Asher, Amsterdam.
Fourie, S. 1962. Notes on a new tritylodontid from the cave sandstone of South Africa. Navors, Natural Museum, Bloemfontein, 2:7-19.
Frets, G. 1909. Über den Plexus lumbosacralis. . . . Morph. Jahrb. Leipzig., 40:1-104.
Furbinger, M. 1874. Zur vergleischenden der schultermuskeln. III. Morph. Jahrb., 1:636-816.

Gadow, H. 1882. Beitrage zur Myologie der hinteren Extremitat der Reptilien. Morph. Jahrb., 7:329-466.
Gadow, H. 1902. Amphibians and Reptiles. Black and Co., London.
Goodrich, E. 1958. Studies on the Structure and Development of Vertebrates, Dover Pub., New York.
Grant, J. 1951. An Atlas of Anatomy. William and Wilkins Co., Baltimore.
Gregory, W., and C. Camp. 1918. Studies in comparative myology and osteology. Bull. Amer. Mus. Nat. Hist., 38:477-563.
Gregory, W. 1947. The monotremes and the palimpset theory. Bull. Amer. Mus. Nat. Hist., 88:7-52.
Griffiths, M. 1968. Echidnas. Pergamon Press, New York.
Gunther, A. 1867. Contributions to the anatomy of Hatteria (Rhyncocephalus, Owen). Phil. Trans. Roy. Soc. London, 157:595-627.

Haines, R. 1934. The homologies of the flexor and adductor muscles of the thigh. J. Morph., 56:21-49.
Haines, R. 1935. A consideration of the constancy of muscular nerve supply. J. Anat., 70:3-55.
Haines, R. 1938. Some musculature changes in the tail and thigh of reptiles and mammals. J. Morph., 58:355-383.
Hartman, C. 1916. Studies in the development of the opossum *Didelphys virginiana*. J. Morph., 27:1-62.
Hopson, J. 1969. The origin and adaptive radiation of mammal-like reptiles and nontherian mammals. Annals New York Academy of Science, 167:199-216.
Hopson, J. 1970. Classification of non-therian mammals. J. Mamm., 51(1):1-9.
Howell, A. 1936. The phylogenetic arrangement of the muscular system. Anat. Rec.,

66:295-316.

Howell, A. 1938. Morphogenesis of the architecture of the hip and thigh. J. Morph., 62:177-218.

Howes, G., and Swinnerton, H. 1900. On the development of the skeleton of the tuatara, *Sphenodon punctatus*. Trans. Zool. Soc., London, 16:B ff.

Huxley, H. 1965. The mechanism of muscular contraction. Scientific American, 213:18-27.

Huxley, J. 1958. Evolutionary processes and taxonomy with special reference to grades. Uppsala University Arsskr., 21-39.

Hyman, L. 1944. Comparative Vertebrate Anatomy. Univ. of Chicago Press, Illinois.

Jenkins, F. 1970a. Cynodont postcranial anatomy and the "prototherian" level of mammalian organization. Evolution, 24:230-252.

Jenkins, F. 1970b. Limb movements in a monotreme (*Tachyglossus aculeatus*) a cineradiographic analysis. Science, 168:1473-1475.

Johnstone, W., and E. Ealey. 1965. Reproduction in the echidna. Australian Mamm. Soc. Bulletin, 2:39.

Kent, G. 1965. Comparative Anatomy of the Vertebrates. The C. V. Mosby Company, St. Louis.

Kermack, D., K. Kermack, and F. Mussett. 1956. New Mesozoic mammals from South Wales. Proc. Geol. Soc. London,#1533:31.

Kermack, K., and F. Mussett. 1958. The jaw and articulation of Docodonta and the classification of Mesozoic mammals. Proc. Royal Society B, 149:204-215.

Kermack, K. 1967. Interrelations of early mammals. J. Linn. Soc. (Zool.), 47:241-249.

Kerr, N. 1955. The homologies and nomenclature of the thigh muscles of the opossum, cat, rabbit. Anat. Rec., 126:481-493.

Klingener, D. 1964. The comparative myology of four dipodoid rodents. Misc. Pub., Museum of Zool., Univ. of Michigan, No. 124, Ann Arbor.

Konisberg, I. 1964. The embryological origin of muscle. Scientific American, August:3-8.

Kolesnikow, W. 1933. Zur vergleichendon Anatomie des M. glutaeo-biceps de Saugetiere. Zeits. f. Anat. u. Entwickl., 99:538-570.

Kühne, W. 1956. The Liassic therapsid oligokyphus. British Museum, London, 10:149.

Kurtén, B. 1968. Pleistocene Mammals of Europe. Weidenfeld and Nicolson, London.

Langworthy, O. 1932. The panniculus carnosus and the pouch of the opossum. J. Mamm., 13:241-251.

Lewis, O. 1961. The phylogeny of the crural and pedal flexor musculature. Proc. Zool. Soc. London, 138:77-109.

Lewis, O. 1963. The monotreme cruro-pedal flexor musculature. J. Anat., London, 97:55-63.

Low, J. 1929. Contributions to the development of the pelvic girdle. The pelvic girdle and its related musculature in monotremes. Proc. Zool. Soc. London, III, 245-265.

Lydeker, W. 1899. Osteology of the Mammalia-Living and Extinct. Black and Co., London.

MacIntyre, G. 1967. Foramen ovale and quasimammals. Evol., 21:834-841.

Maurer, F. 1898. Die Entwickling der Ventralen Rumpf Muscularus bei Reptilien. Morphologisches Jahrbuch, 26:1-60.

Mayr, E. 1963. Animal Species and Evolution. Harvard Univ. Press., Cambridge.

McCrady, E. 1938. The Embryology of the Opossum. Amer. Anat. Memoirs, No. 16, 1-234.

Mivart, S. 1866. On some points in the anatomy of *Echidna hystrix*. Trans. Linn. Soc., 25:379-402.

Nairin, R., and W. DeBoer. 1966. Species distribution of gastro-intestinal antigens. Nature, 210:960-962.

Napier, J. 1967. The antiquity of human walking. Scientific Amer., 216:(4)56-66.

Nelson, O. 1953. Comparative Embryology of the Vertebrates. McGraw-Hill Book Co., Inc., New York.

Newman, A. 1877. Notes on the physiology and anatomy of the tuatara. Tr. New Zealand Institute, 10:222-239.

Noble, G., and E. Noble. 1939. A Brief Anatomy of the Turtle. Stamford Univ. Press, California.

Nomina Anatomica. 1965. (2nd ed., 1968). International Congress of Anatomists. Excerpta Medica Foundation, Amsterdam.

Nomina Anatomica Verterinaria. 1968. International Committee on Veterinary Anatomical Nomenclature. World Assoc. of Veterinary Anatomists, Vienna.

Olson, E. 1958. The evolution of mammalian characteristics. Evol., 13:344-353.

Osawa, G. 1898. Beitrage zur Anatomie der *Hatteria punctata*. Arch. f. Anat., 51:481-491.

Osgood, W. 1921. A Monographic Study of the American Marsupial, Caenolestes. Field Mus. Nat. Hist., Chicago, 14:3-156.

Owen, R. 1866. Anatomy and Physiology of Vertebrates. 1, 2, 3, Longmans, Green & Co., London.

Owen, R. 1867. On the fossil mammals of Australia. Part III Phil. Trans. Lond., 160:519-578.

Parrington, F. 1961. The evolution of the mammalian femur. Proc. Zool. Soc. London, 137:285-298.

Parrington, F. 1967. The origins of mammals. Presidential address. Section D (Zoology). British Association for Advancement of Science, 165-173.

Paterson, A. 1892. The pectineus muscle and its nerve supply. J. Anat. Physiol., 26:43-47.

Pearson, H. 1926. Pelvic and thigh muscles of *Ornithorhynchus*. J. Anat., London, 210:152-163.

Perrin, A. 1895. Researches sur les affinités zoologiques de l'*Hatteria punctata*. Annales des Sciences Naturelles, Zool. et Pal., 20:33-102.

Poluhowich, J., and A. Brush. 1971. An electrophoretic study of *Sphenodon* proteins. (In Press).

Pournelli, G. 1957. The echidna a mammalian throwback. Zoo. Mag., 30(6):116-120.

Rabl, C. 1916. Uber der Muskeln und Nerven der Extremitaten von *Iquana tuberculata*, Gray. Anat. Hefte, 58:681-789.

Reed, C. 1960. Polyphyletic or monophyletic ancestry of mammals. Evolution, 14:314-22.

Reighard, J., and H. Jennings. 1963. Anatomy of the Cat (rev., 3rd ed.). Holt, Rinehart and Winston, New York.

Ride, W. 1964. A review of Australian fossil marsupials. Journal Royal Society W. Australia, 47:97-131.

Romer, A. 1922. The locomotor apparatus of certain primitive and mammal-like reptiles. Bull. Amer. Mus. of Nat. Hist., 46:517-603.

Romer, A. 1923. Crocodilian pelvic muscles and their avian and reptilian homologs. Bull. Amer. Mus. Nat. Hist., 48:533-552.

Romer, A. 1942. The development of the limb musculature of tetrapods: the thigh of Lacerta. J. Morph., 71:251-98.

Romer, A. 1964. The Vertebrate Story. University of Chicago Press. Chicago.

Romer, A. 1966. Vertebrate Paleontology. The Univ. of Chicago Press, Chicago.

Romer, A. 1967a. The Chañares (Argentina) Triassic reptile fauna. Mus. of Comp. Zool., Breviora, #264.

Romer, A. 1967b. Early reptilian evolution reviewed. Evolution, 21:821-833.

Romer, A. 1968a. Osteology of the Reptiles. The Univ. of Chicago Press, Chicago.
Romer, A. 1968b. Notes and Comments of Vertebrate Paleontology. Univ. of Chicago Press, Chicago.
Romer, A. 1970. The Vertebrate Body (4th ed.). W. B. Saunders Co., Philadelphia.
Romer, A., and L. Price. 1940. A review of the pelycosaur. Geol. Soc. Amer. Spec. Papers, 28.
Romer, F. 1898. Des Integument der Monotremen, Denkschr. D. Med. Naturwiss Gesellsch., Jena. 6:189-242.

Schauinsland, H. 1900. Weitere Beitrage zur Entwicklingsgeschichte der Hatterica. Archiv. f. mikrosk. Anat., 56:747-867.
Semon, R. 1894a. Beobachtungen ueber die Lebensweise und Fortpflanzung der Monotremen. Denkschr. Med. Naturwiss Gesellsch., Jena, 5:3-15.
Semon, R. 1894b. Beschreibung der Embryonalhullen der Monotremen und Marsupialier. Denkschr. Med. Naturwiss Gesellsch., Jena, 5:19-58.
Sharell, R. 1966. The Tuatara, Lizards, and Frogs of New Zealand. Collins, London.
Simpson, G. 1953. The Major Features of Evolution. Yale Univ. Press, New Haven.
Simpson, G. 1959. Mesozoic mammals and the polyphyletic origin of mammals. Evolution, 13:405-14.
Simpson, G. 1960. Diagnosis of the classes reptilia and mammalia. Evolution, 14:388-92.
Simpson, G. 1961. Historical zoogeography of Australian mammals. Evolution, 15:431-446.
Simpson, G. 1964. The Meaning of Evolution. Yale Univ. Press, New Haven.
Simpson, G. 1969. Principles of Animal Taxonomy. Columbia University Press, New York.
Sissons, S. 1964. Anatomy of Domestic Animals. W. B. Saunders Co., Philadelphia.
Sonntag, C. 1920. Contributions to the visceral anatomy and myology of the marsupalia. Proc. Zool. Soc. London, 851-852.

Van Valen, L. 1960. Therapsids as mammals. Evolution, 14:304-13.
Vaughn, P. 1956. The phylogenetic migration of the ambiens muscle. J. Elisha Mitchell Scientific Society, 72:243-262.
Voss, H. 1963. Besitzen die Monotremen (*Echidna* and *Ornithorhynchus*) ein-oder mehrfaserige Muskelspindeln? Anat. Anz., 113:255-258.

Walker, F. 1965. Mammals of the World. The Johns Hopkins Press, Baltimore.
Waring, H., R. Noir, and C. Tyndale-Biscoe. Comparative physiology of marsupials. Advances in Comparative Physiology and Biochemistry (ed. Lowenstein). 2, Academic Press, N.Y.
Watson, D., and A. Romer. 1956. A classification of the therapsid reptiles. Bull. Mus. Comp. Zool., 114:35-89.
Westling, C. 1889. Anatomische Untersuchungen uber Echidna. Behang. J. Svenska Vet-Akad. Handlinjar, 15:377-388.
Wemyss, C., 1943. A preliminary study of marsupial relationships as indicated by the protein precipitin test. Zoologia. Dec.
Wheeler, R. 1934. The homologies of the flexor and adductor muscles of the thigh. J. Morph., 56:21-49.
Wheeler, R. 1938. Some muscular changes in the tail and thigh of reptiles and mammals. J. Morph., 58:355-383.
Williston, S. 1925. Osteology of the Reptiles. Harvard Univ. Press, Cambridge.
Woods, J. 1962. Fossil marsupial and Cainozoic continental stratigraphy in Australia: a review, Mem. Queensland Museum, 14:41-49.

Young, J. 1950. Life of Vertebrates. Oxford Press, New York.

PLATE 1
DIDELPHIS — VENTRAL

a, Sartorius
b, Rectus femoris
c, Vastus 3
d, Vastus 4
e, Adductor 1
f, Adductor 2
g, Adductor 3
h, Semimembranosus
(2 bundles)
i, Iliopsoas
j, Gracilis
k, Glans ani
l, Pyramidalis
m, Os marsupialis (Area)
n, Fascia (Iliacus)
o, Arteria femoralis
Vena femoralis
p, Patella
q, Crus (leg)
r, Semitendinosus
s, N. obturatorius
(Arrow to Gracilis)
t, Locus-Pectineus
u, Crus ligamentum
v, Adductor 4 (fused
with Adductor 3)
w, Iliacus (Area)

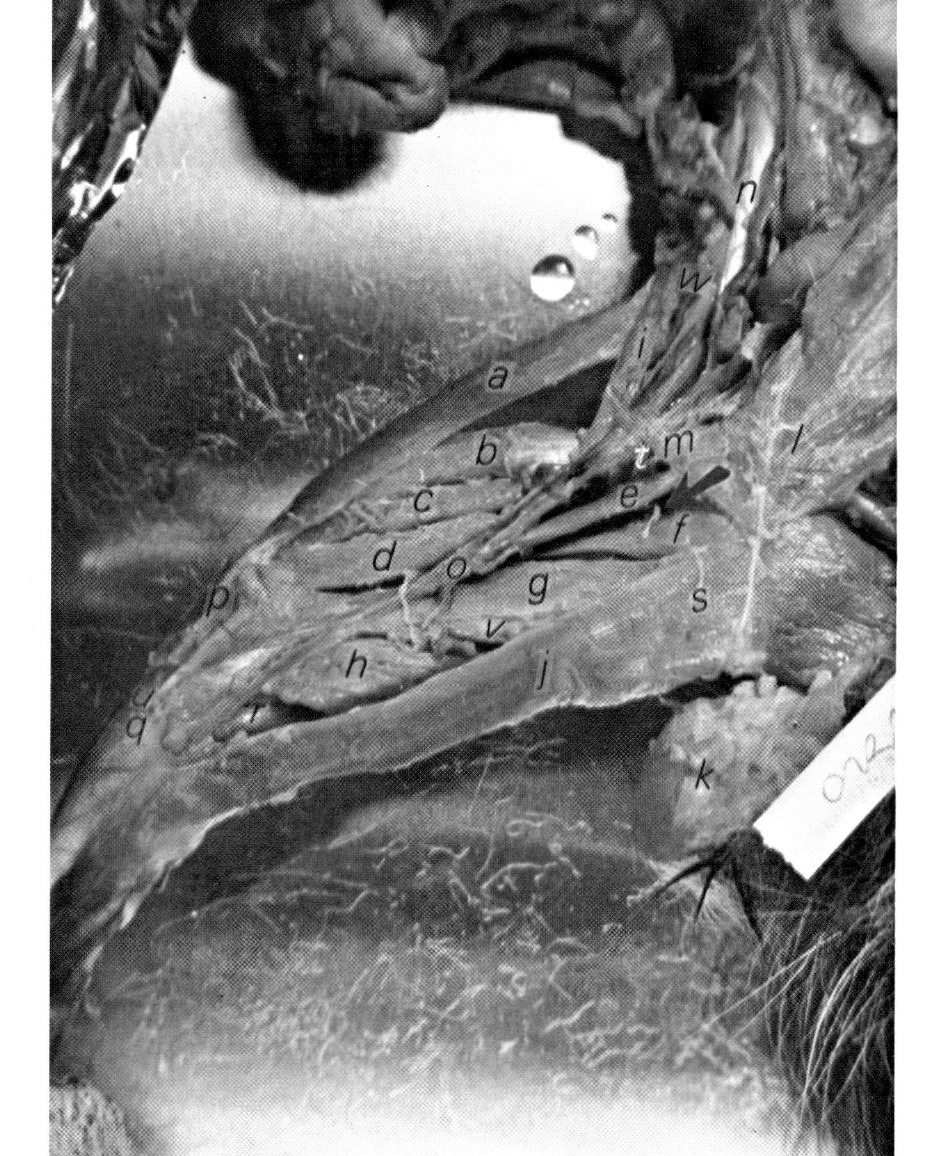

Right hind limb of the opossum, Didelphis, showing superficial muscles in ventral view. N. pyramidalis is bisected (arrow). Magnification close to actual size.

PLATE 2

TACHYGLOSSUS — VENTRAL

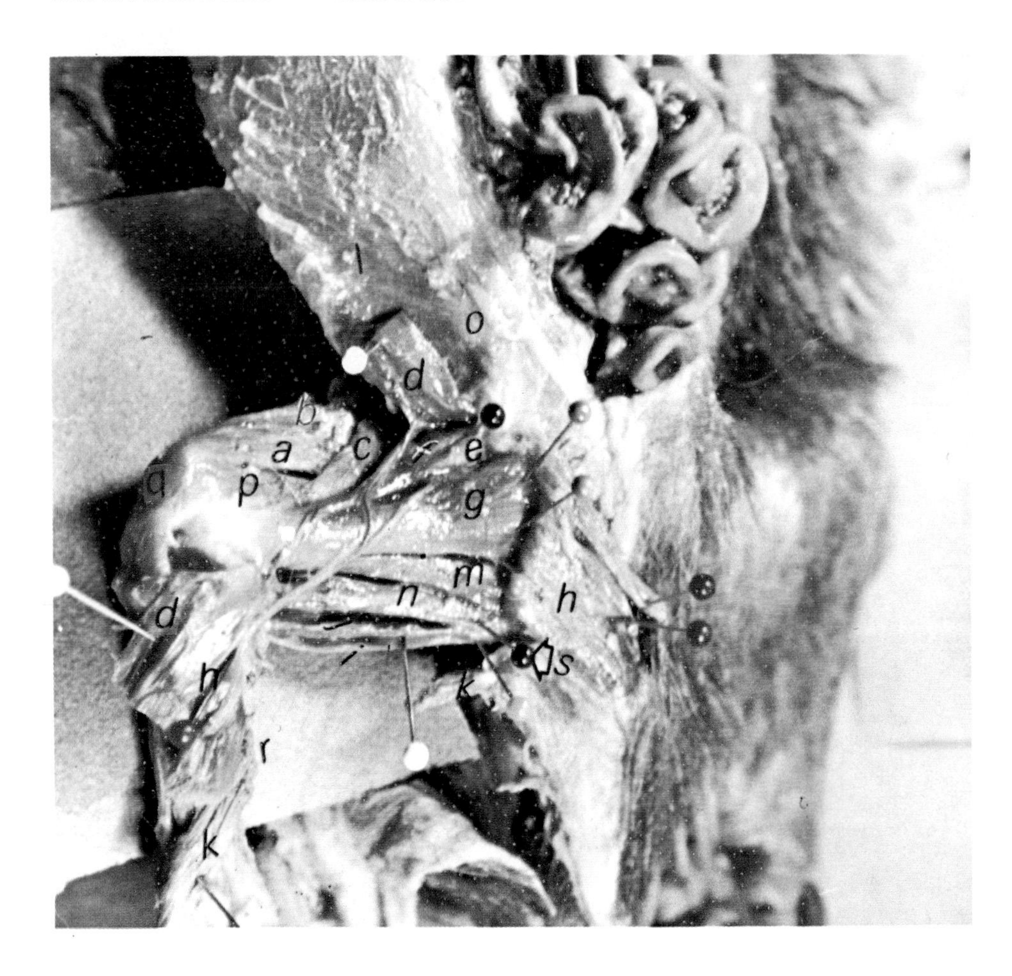

a, Vastus 3
b, Rectus femoris
c, Iliopsoas (Iliacus-Psoas)
d, Sartorius
e, Adductor 1
f, Pectineus (2 muscles)
g, Adductors 2 & 3
h, Gracilis (2 muscles)
i, Biceps femoris (fused with semitendinosus, j)
j, Semitendinosus (dorsal head)
k, Biceps femoris
l, Obliquus abdominis externus
m, Semimembranosus (fusion of 2 or 3 muscles)
n, Semitendinosus (ventral head)
o, Os marsupialis (Area)
p, Vastus 4
q, Patella
r, Crus (leg)
s, Semitendinosus (smaller head of n — arrow)

Right hind limb of an immature echidna, Tachyglossus, revealing superficial muscles in ventral view. The most superficial muscles are bisected at belly and reflected; (Mm. sartorius, d, gracilis, h, biceps femoris, k).

PLATE 3
DIDELPHIS – DORSAL

a, Semitendinosus (Crurococcygeus)
b, Semimembranosus (Fused area)
c, Biceps femoris
d, Femorococcygeus
(Femorocaudalis)
e, Gastrocnemius
f, Vastus 1
g, Gluteus 2
h, Sartorius
i, Gluteus 3
j, Gluteus 1
k, Glans ani
l, Transversospinalis
system
m, Longissimus
lumborium
n, Latissimus dorsi
o, Rectus femoris
p, Iliocostalis
thoracis
q, Longissimus
thoracis
r, Semimembranosus
(divides)
s, Semitendinosus
t, Presemimembranosus
u, Semimembranosus
v, Vastus 2

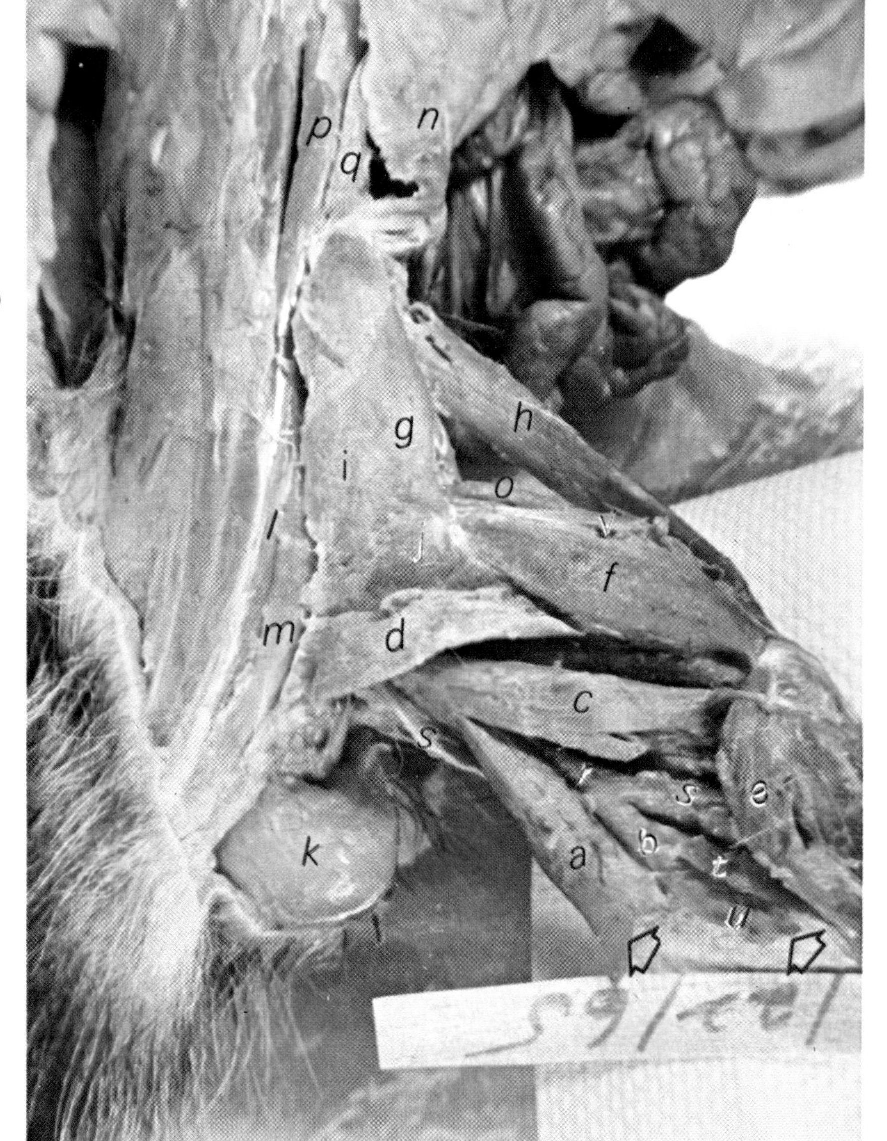

Right hind limb of the opossum, Didelphis, showing the superficial muscles in dorsal view. Muscles in which a muscle bundle was cut are indicated by arrows. Magnification close to actual size.

PLATE 4

TACHYGLOSSUS — DORSAL

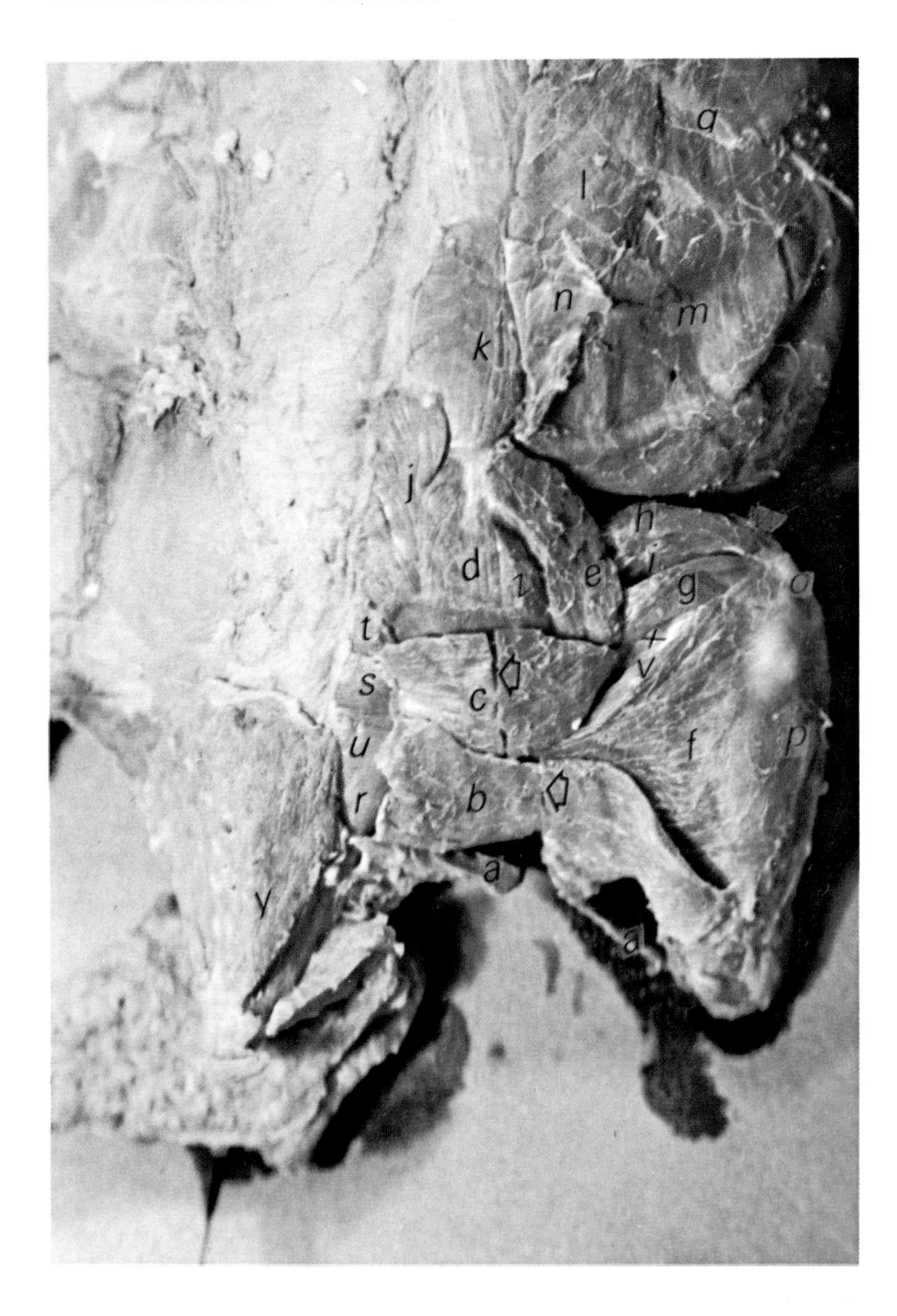

a, Femorococcygeus (--caudalis; caudal part of b)
b, Crurococcygeus (--caudalis; fragmented slip)
c, Gluteus 1
d, Gluteus 3
e, Gluteus 2
f, Biceps femoris
g, Vastus 1
h, Vastus 2
i, Vastus 3
j, Gluteus 4
k, Erector spinae (Fused Transversospinalis and Longissimus)
l, Latissimus dorsi
m, Obliquus abdominis externus
n, Thorax-lumbricus fascia (reflected)
o, Patella (Area)
p, Crus (leg)
q, Costae thoracis
r, Sacrocaudalis dorsalis
s, Piriformis
t, Gluteus 5
u, Coccygeus
v, Biceps femoris (partim)
w, Rectus femoris
x, Gastrocnemius
y, Cutaneous muscle (overlaying vertebrae)
z, Gluteus 6

Right hind limb of the echidna, Tachyglossus, showing the superficial muscles in dorsal view (the pes has been removed). Bisected muscle bundles are indicated (arrows). Magnification about actual size.

PLATE 4A
TACHYGLOSSUS – LEFT DORSAL

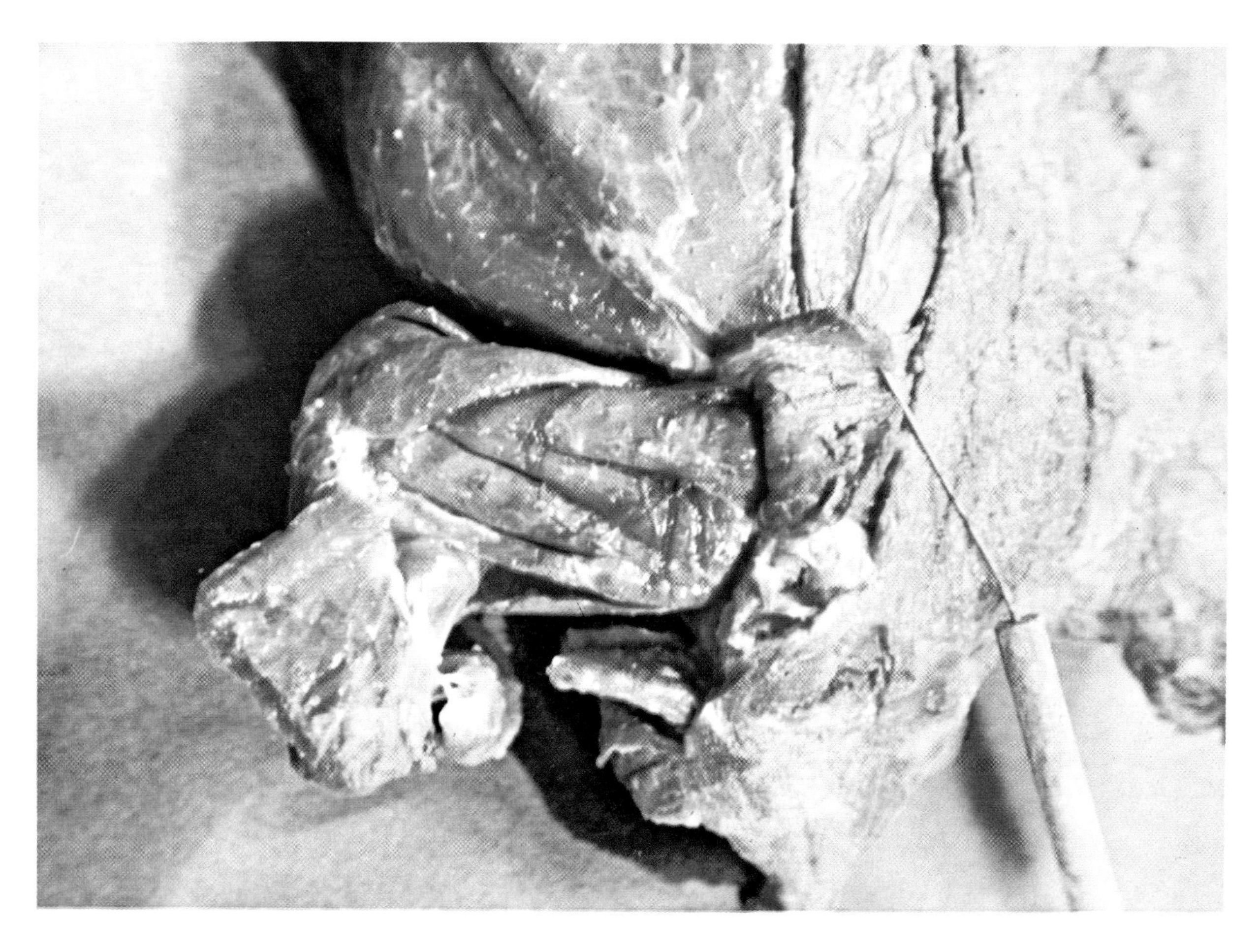

Tachyglossus, left dorsal view, superficial muscles reflected to show mesial bundles.

PLATE 5
SPHENODON — VENTRAL

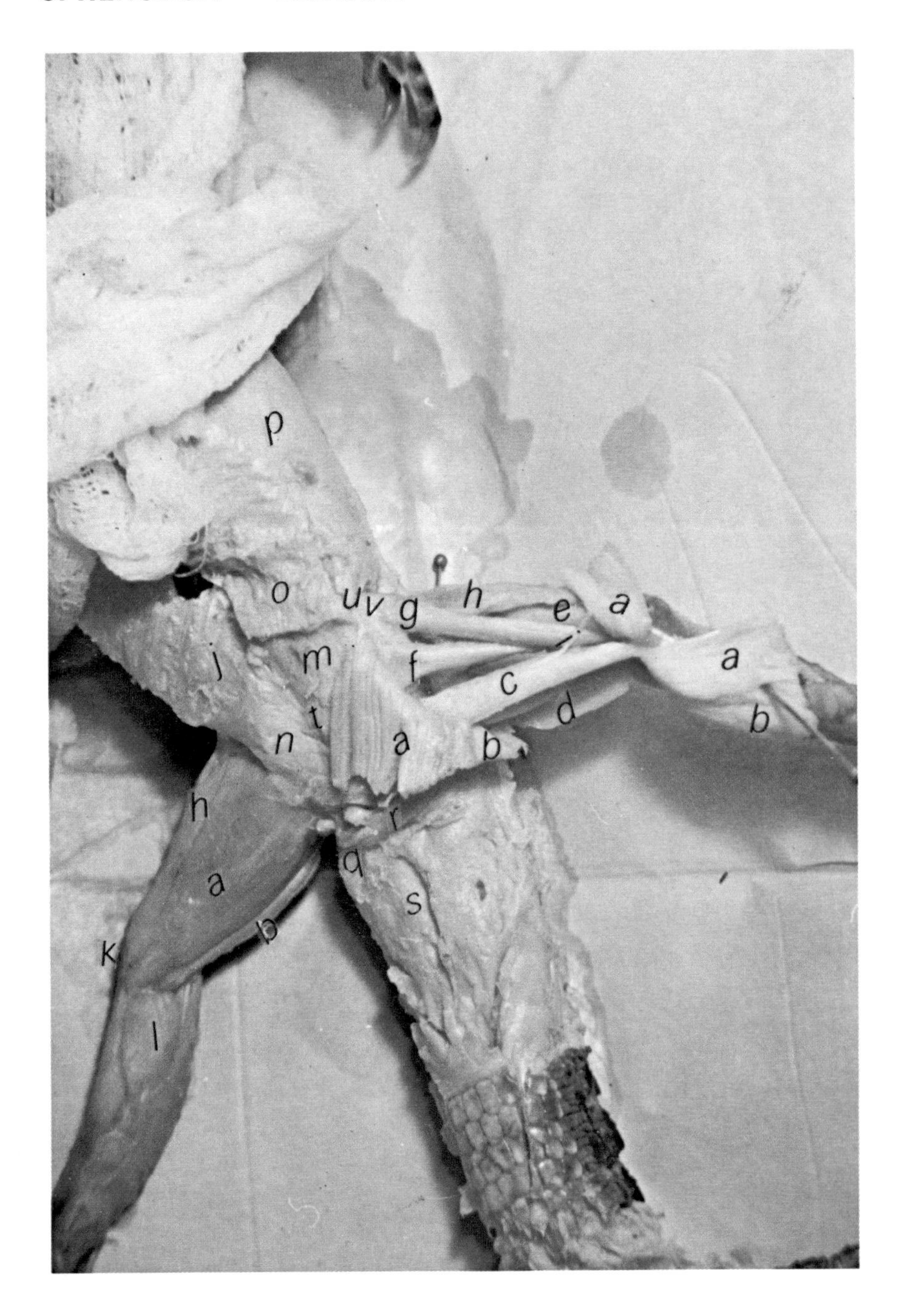

a, Gracilis
b, Semitendinosus
c, Adductor 1
d, Semimembranosus
e, Vastus (divides)
f, Adductor 2
g, Adductor 3
h, Sartorius
i, N. obturatorius
j, Obliquus abdominis externus (cut)
k, Articularis genu
l, Gastrocnemius (divides)
m, Obliquus abdominis internus
n, Fascia of m (above)
o, Cut of j (above) to show m
p, Latissimus dorsi
q, Femorococcygeus (Femorocaudalis)
r, Anus externus
s, Sacrocaudalis ventralis medialis
t, Symphysis
u, Locus-Prepubic tubercle
v, Iliopsoasp-pectineus

Right hind limb of a female tuatara, Sphenodon, showing superficial muscles in ventral view. The Mm. gracilis, a, semimembranosus, d, and semitendinosus, b, are bisected and reflected. 2X.

PLATE 6

SPHENODON — DORSAL VIEW

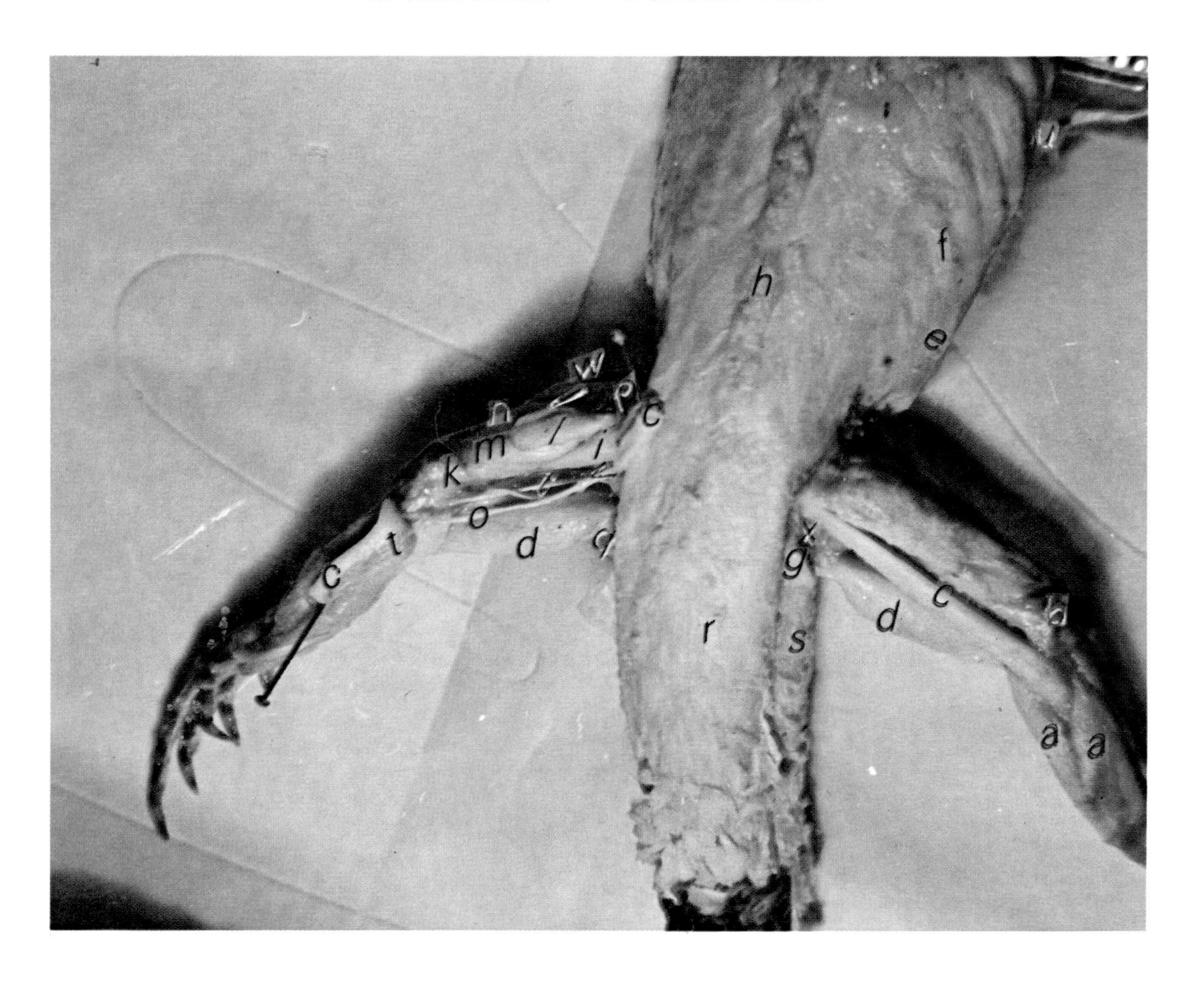

Right and left hind limbs of a female tuatara, <u>Sphenodon</u>, showing exterior muscles in dorsal view. The M. biceps femoris, c, is reflected on the left crus. 2X.

a, Gastrocnemius
b, Articularis genu
c, Biceps femoris
d, Semimembranosus
e, Obliquus abdominis externus
f, Lumbar ribs within muscle
g, Semimembranosus — semitendinosus sheet
h, Vertebrae lumbalis
i, Gluteus
j, Locus-Vastus (division)
k, Vastus 1
l, Rectus femoris
m, Vastus 2
n, Vastus (division)
o, Plexus sacralis
p, Locus-Mm. Iliopsoas-pectineus
q, Semitendinosus
r, Sacrocaudalis dorsalis medialis
s, Caudae dorsalis lateralis
t, Plexus sacralis innervating c
u, Coxa ventralis (Part of Gastralia)
v, N. Plexus sacralis
w, Sartorius
x, Femorococcygeus (Femorocaudalis)

PLATE 7
SPHENODON — EXTERNAL VIEW
(LIVING SPECIMEN)
OSTEOLOGY

PLATE 8

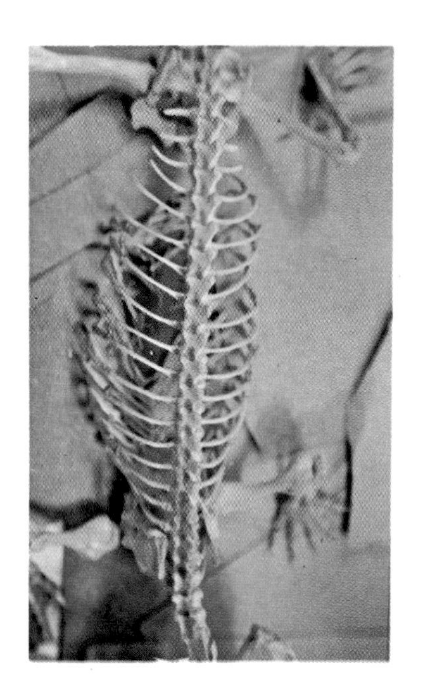

PLATE 9a

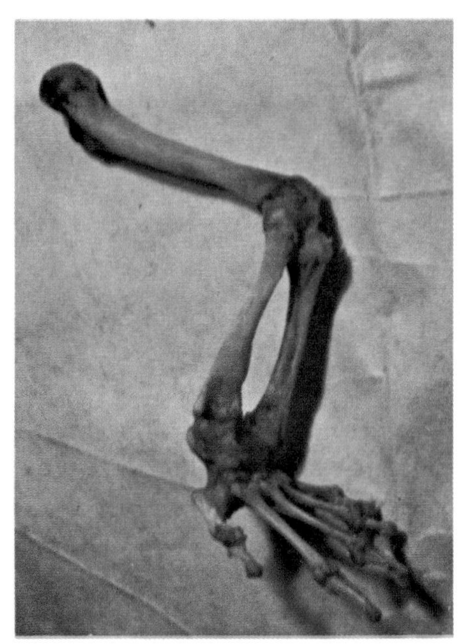

PLATE 9

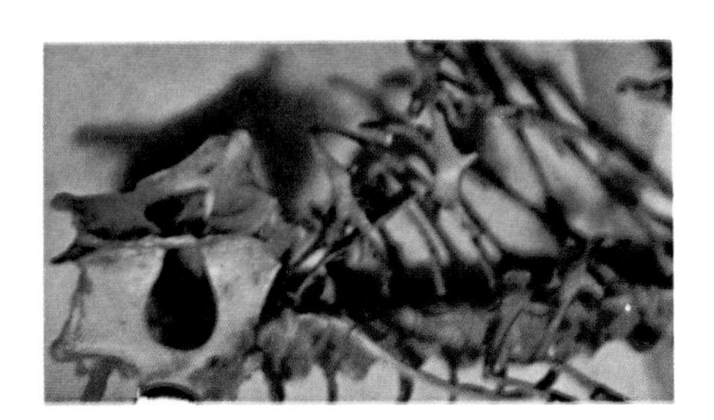

PLATE 10
TACHYGLOSSUS — EXTERNAL VIEW
(PRESERVED SPECIMEN)

OSTEOLOGY

PLATE 10a

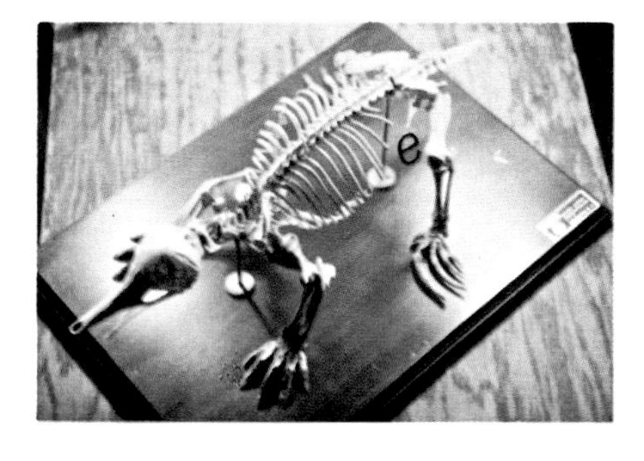

PLATE 10b

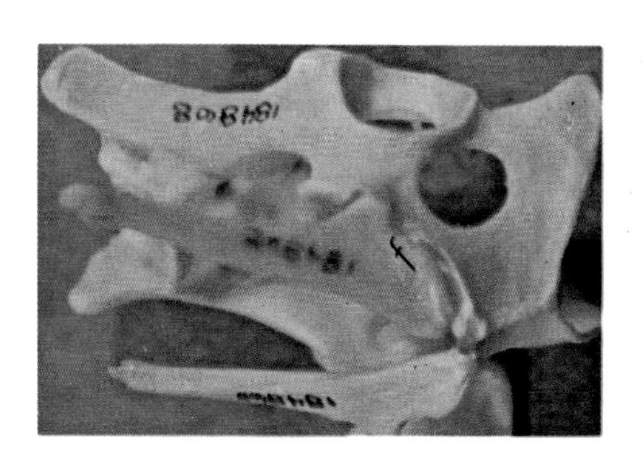

PLATE 10c

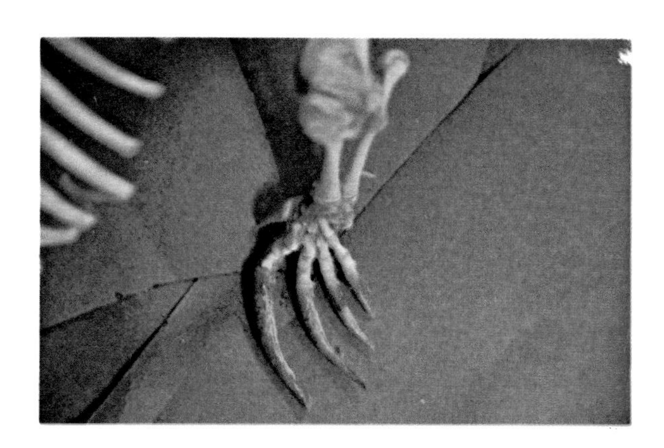

PLATE 11
DIDELPHIS — EXTERNAL VIEW
(PRESERVED SPECIMEN)

OSTEOLOGY

PLATE 12

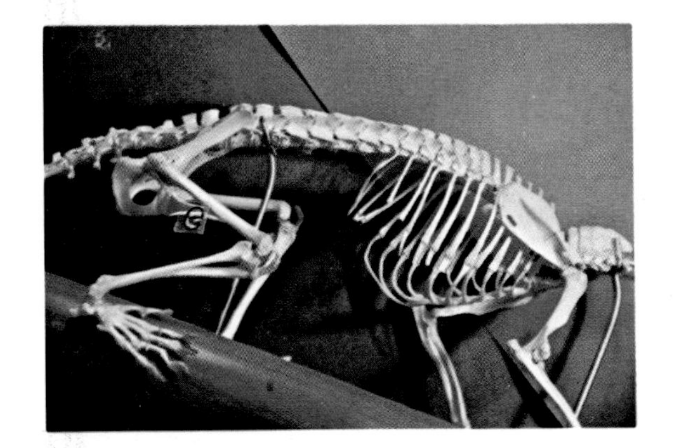

PLATE 12a

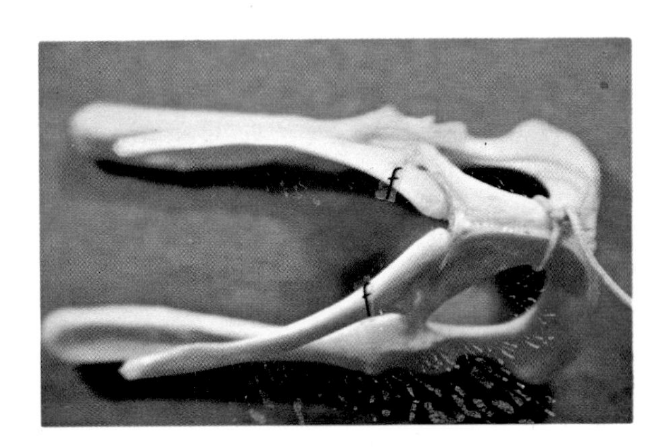

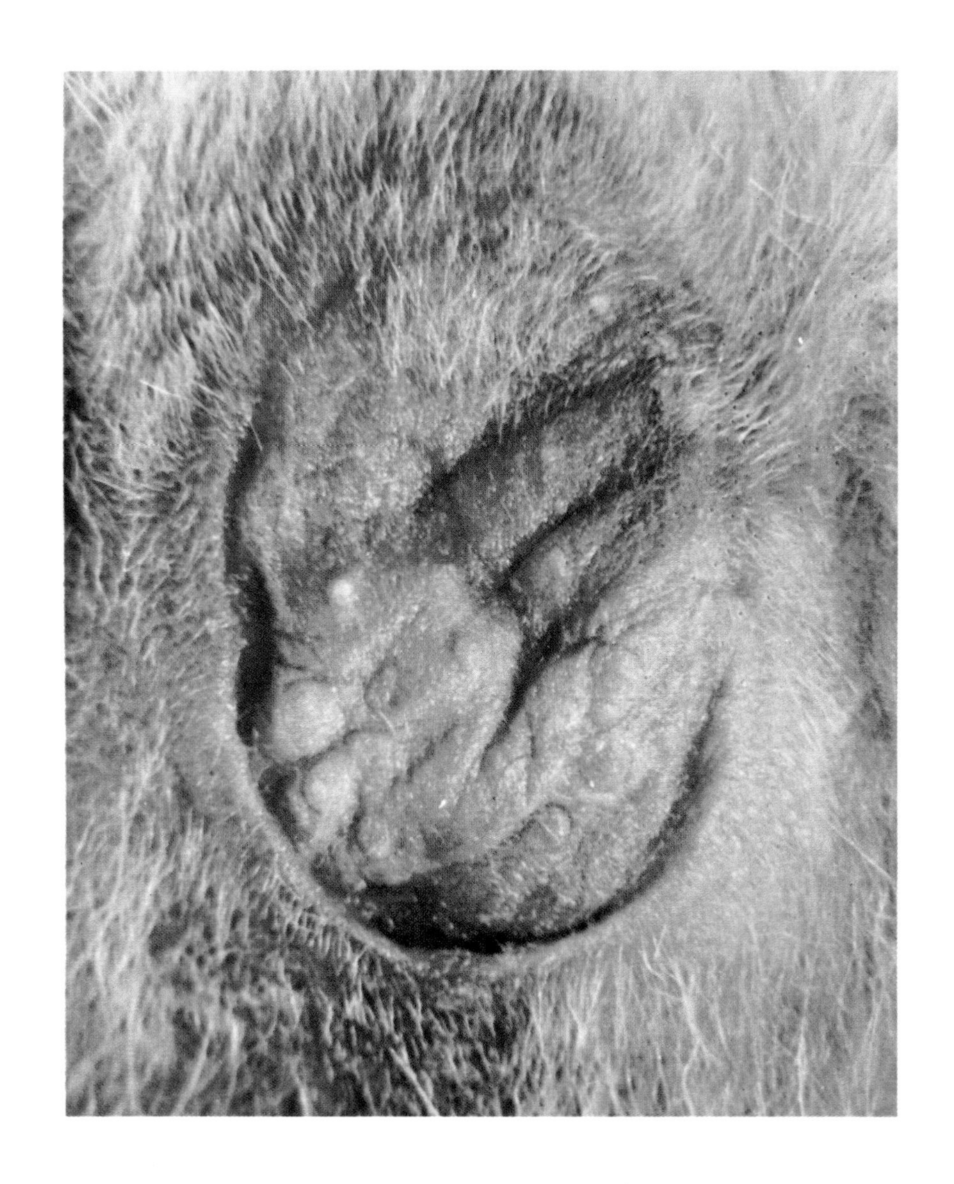

PLATE 12b

MARSUPIUM
OF DIDELPHIS
(NOTE TEATS)

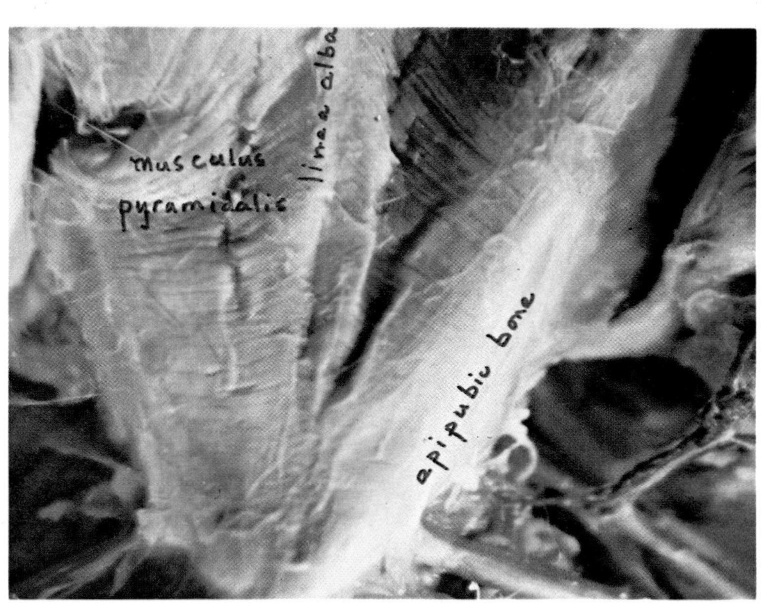

PLATE 12c

EPIPUBIC BONES
LYING BENEATH
AND NEAR
MUSCLUS PYRAMIDALIS
IN THE OPOSSUM, DIDELPHIS

PLATE 12D

MUSCULUS RECTUS ABDOMINIS ABOVE THE MUSCULUS PYRAMIDALIS; tip of epipubic bone at arrow in DIDELPHIS. Musculus obiliquus externus at extreme right.

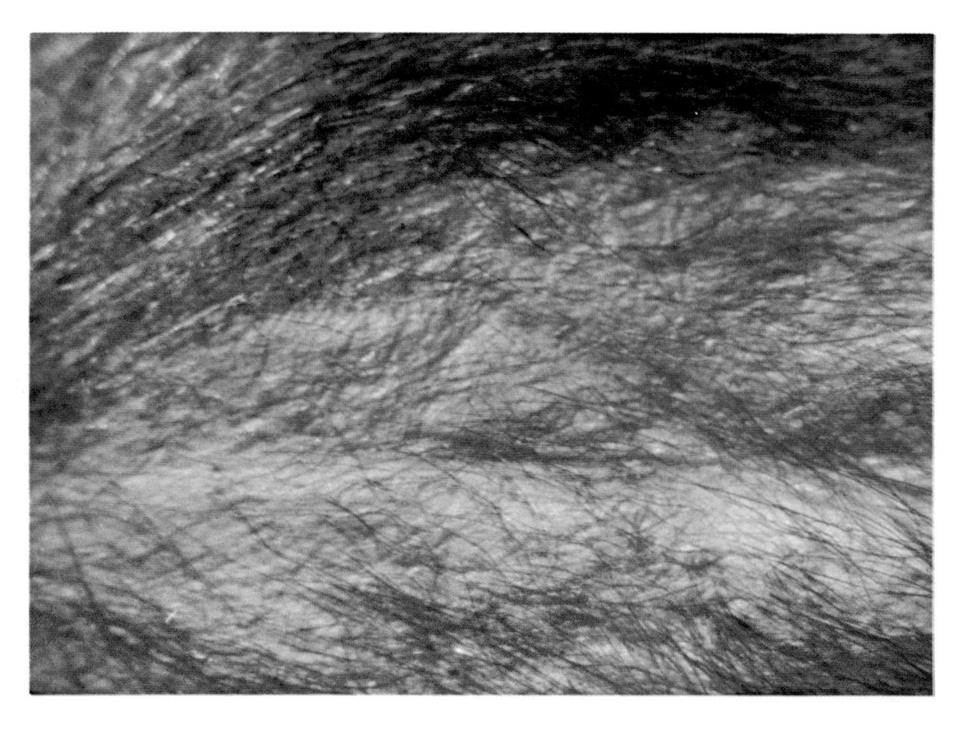

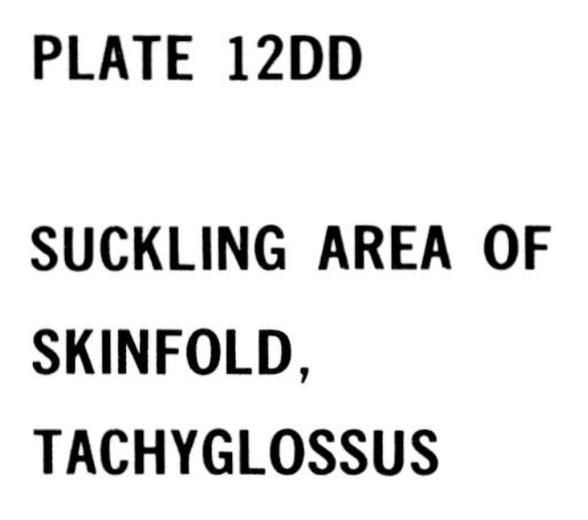

PLATE 12DD

SUCKLING AREA OF SKINFOLD, TACHYGLOSSUS

PLATE 12E

EPIPUBIC BONES IN THE ECHIDNA, TACHYGLOSSUS, Lying wholly below the Musculus obliquus externus (see arrow). M. Pyramidalis, a. (see small plate)

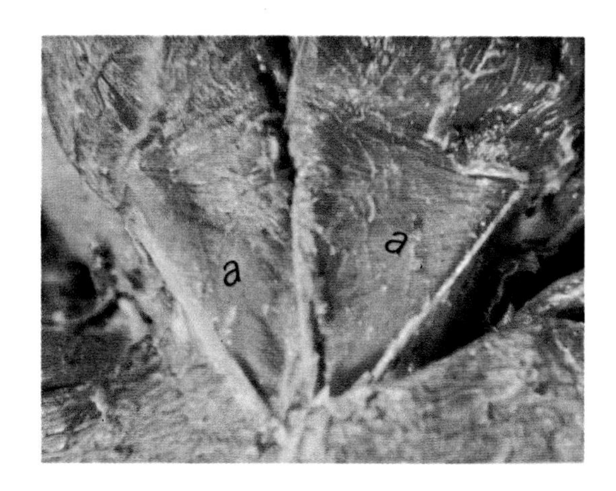

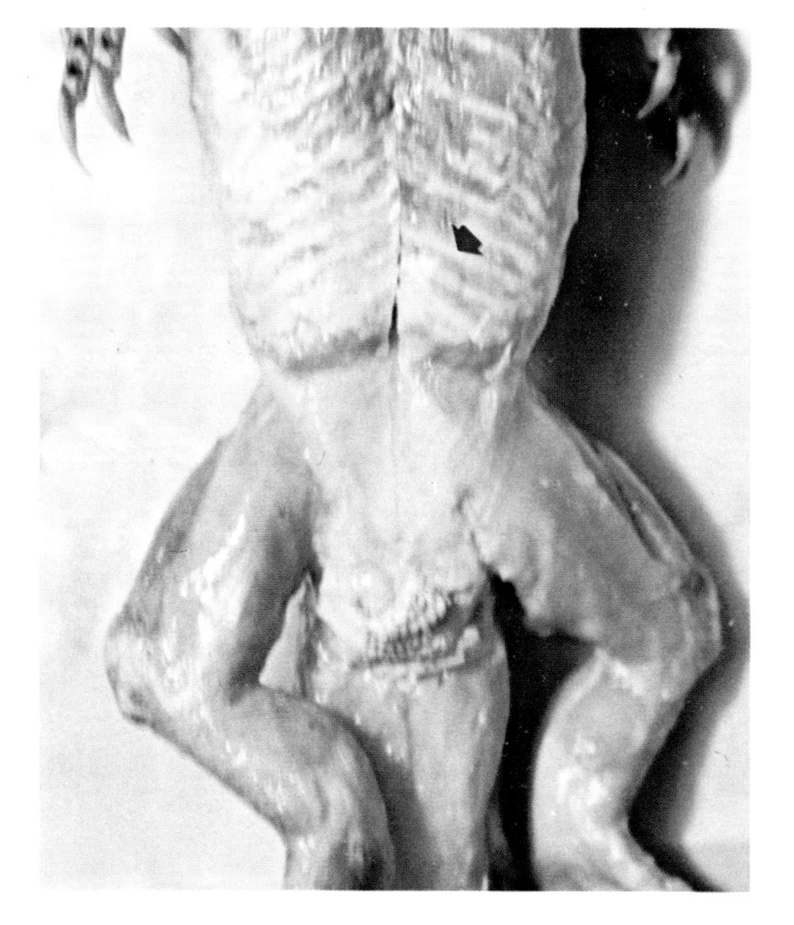

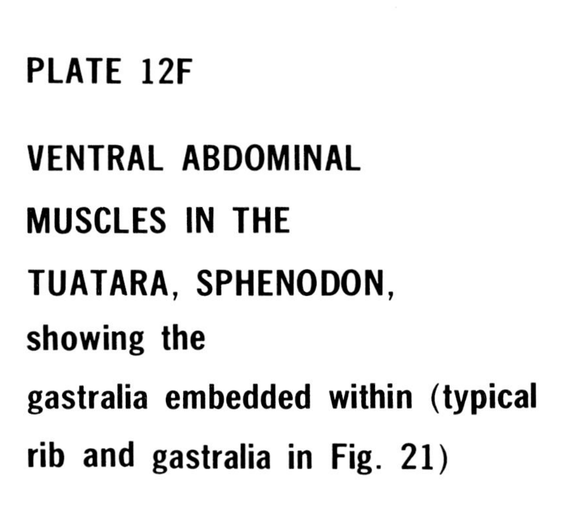

PLATE 12F

VENTRAL ABDOMINAL MUSCLES IN THE TUATARA, SPHENODON, showing the gastralia embedded within (typical rib and gastralia in Fig. 21)

SPHENODON
PLATES 13-21
Dissection of Ventral Pelvis, male and female (X's 2)

PLATE 13

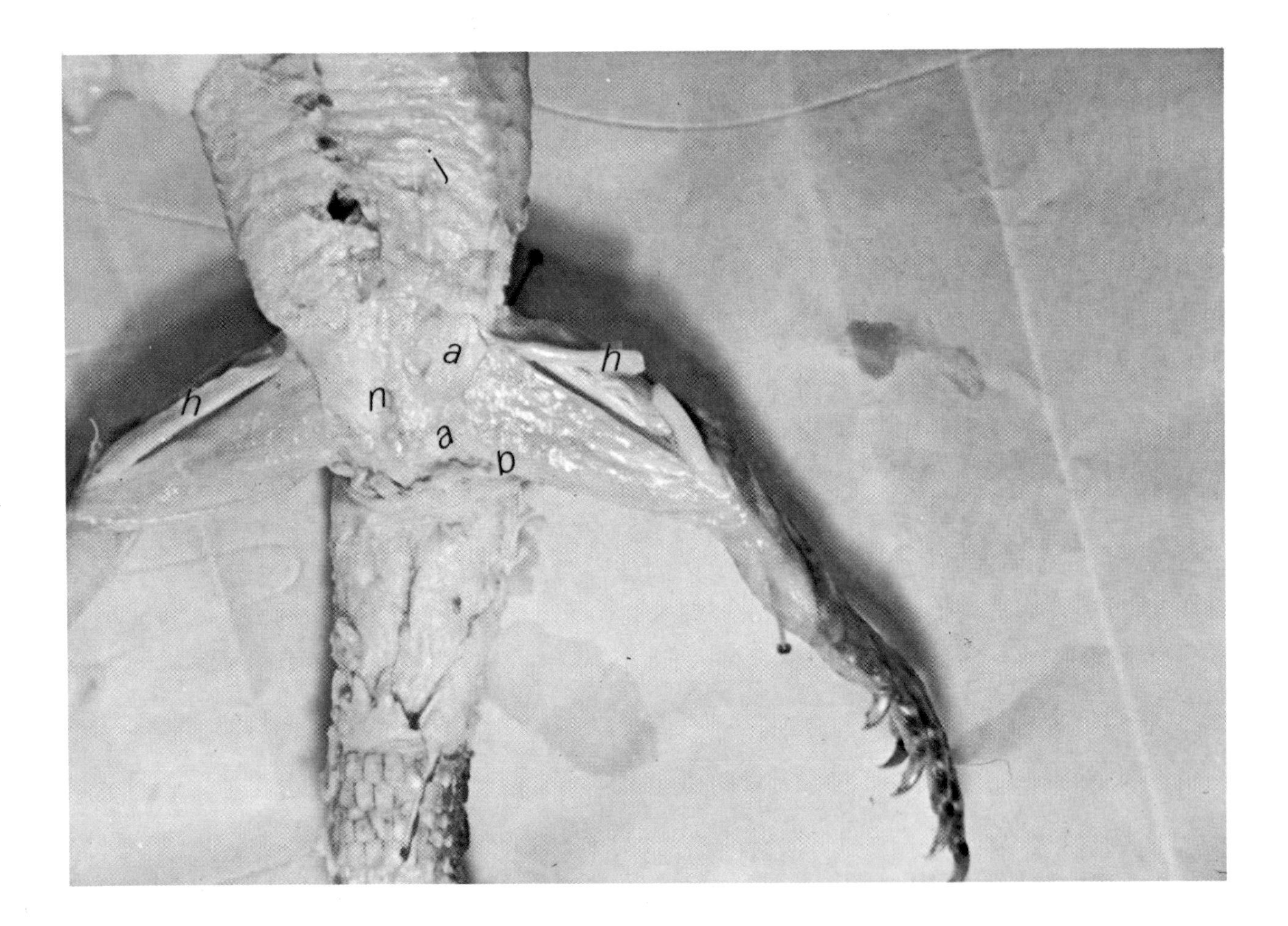

Connective tissue, n. outlines the ventral pelvic plate.

Extensive gracilis, a, originates from ischio-pubic symphysis.

Semitendinosus fibers, b, are attached to gracilis.

Gastralia lie within obliquus abdominis externus, j, anteriorly, just above the girdle.

Sartorius, h, is usual position.

PLATES 14 and 15
EXPLANATION OF FIGURES

Right hind limb of a female tuatara, Sphenodon, showing superficial muscles in ventral view. The Mm. gracilis, a, semimembranosus, d, and semitendinosus, b, are bisected and reflected. 2X.

PLATE 14

a, Gracilis
b, Semitendinosus
c, Adductor 1
d, Semimembranosus
e, Vastus (divides)
f, Adductor 2
g, Adductor 3
h, Sartorius
i, N. obturatorius
j, Obliquus abdominis
  externus
k, Articularis genu
l, Gastrocnemius (divides)
m, Obliquus abdominis
  internus

n, Fascia of m (above)
o, Cut of j (ablve to
  show m
p, Latissimus dorsi
q, Femorococcygeus
  (Femorocaudalis)
r, Anus externus
s, Sacrocaudalis ventralis
  medialis
t, Symphysis
u, Locus-Prepubic tubercle
v, Iliopsoas-pectineus
w, Ventral pelvic plate
  area-obturator externus,
  quadratus femoris, both
  below

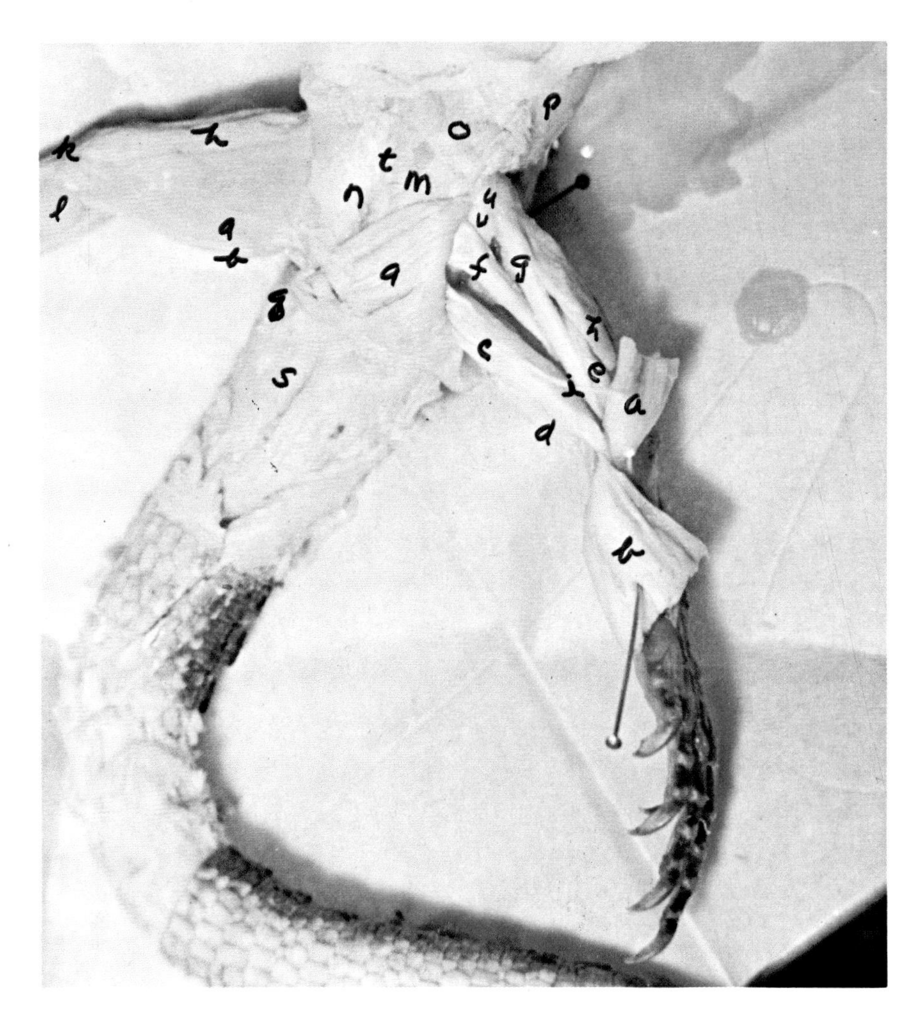

Gracilis, a, bisected at the top.

Gracilis, semitendinosus, b, cut, -reflected.

Nervus obturatorius innervation, i (Plate 14).

PLATE 15

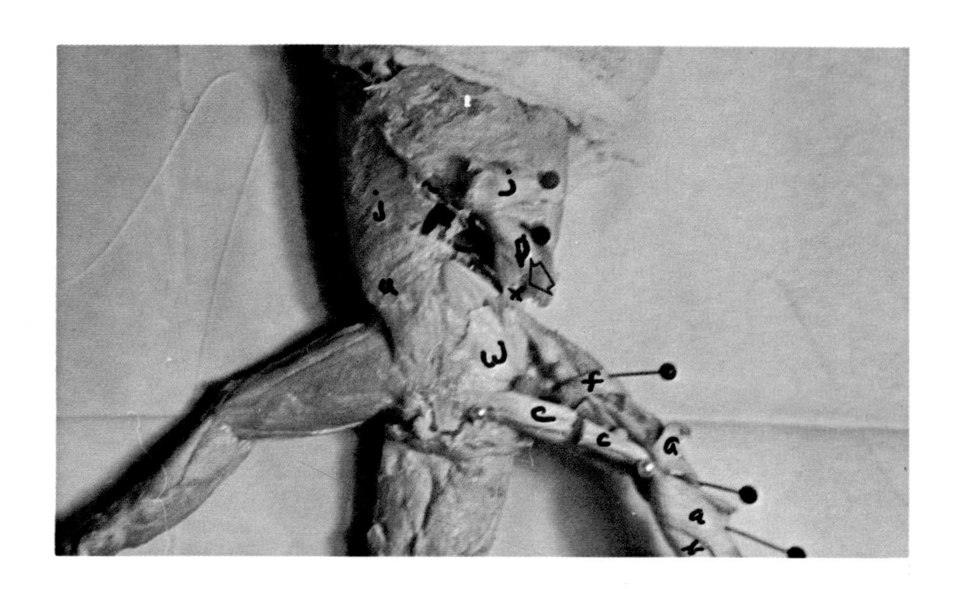

Obliquus abdominis internus, p, can be seen attached to the prepubic tubercle, u; at x,-arrow (Plate 15).

Obturator externus and quadratus femoris on ventral pelvic plate, w (Plates 15; 15b).

Adductores, c, f, and g (f reflected in Plate 15b; g reflected in Plate 15a).

PLATE 15a

PLATE 15b

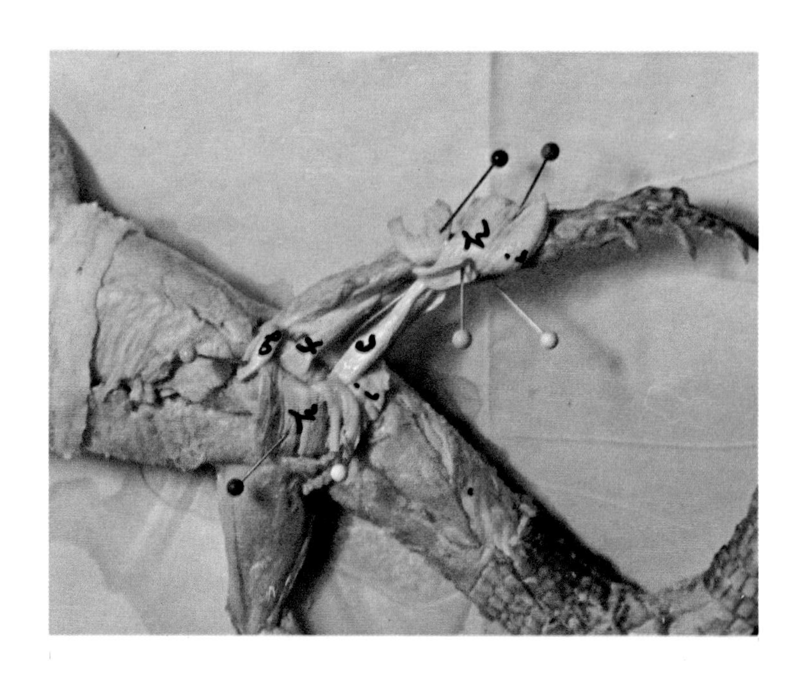

PLATE 16

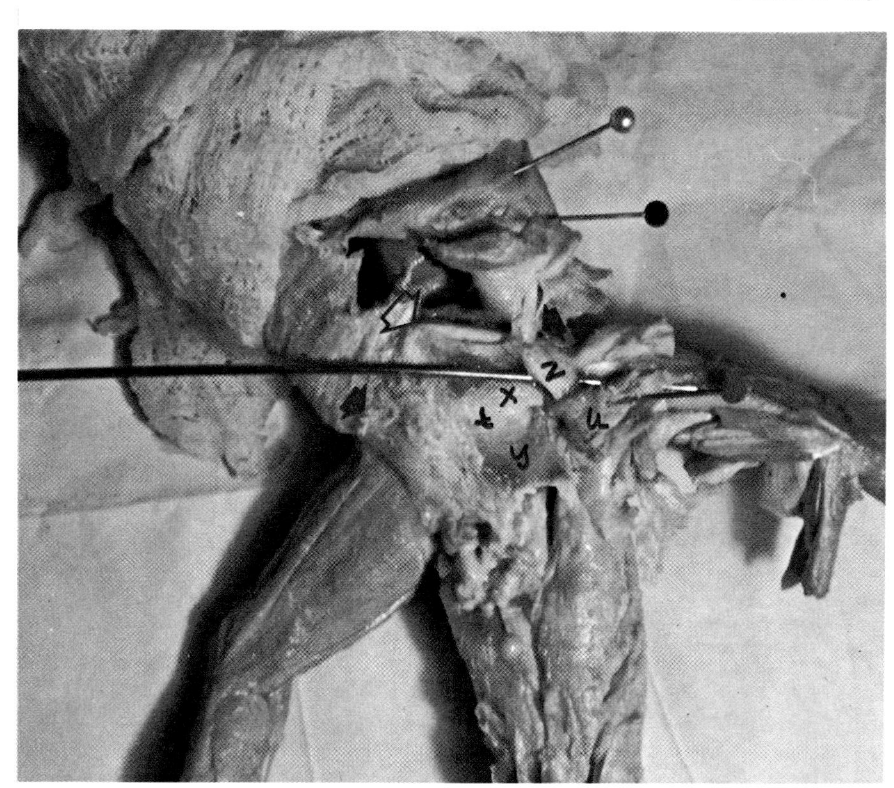

Ventral pelvic plate, y.

Obturator externus, x.

Iliopsoas-pectineus, z.

Adductores, u, cut at symphysis, reflected, Plate 16.

Area of prepubis (black arrows).

Obturator foramina opening covered with ventral fascia, t.

Area of epipubis (white arrow).

Obliquus abdominis externus and internus, reflected-pearly pin.

Bisected half of gracilis, adductores, 1 & 2, removed from cartilagenous symphysis of the ischiopubic plate, reflected-black pin.

Iliopsoas-pectineus bisected-yellow pin, z, Plate 17.

Lower half of adductores removed in Plate 17.

PLATE 17

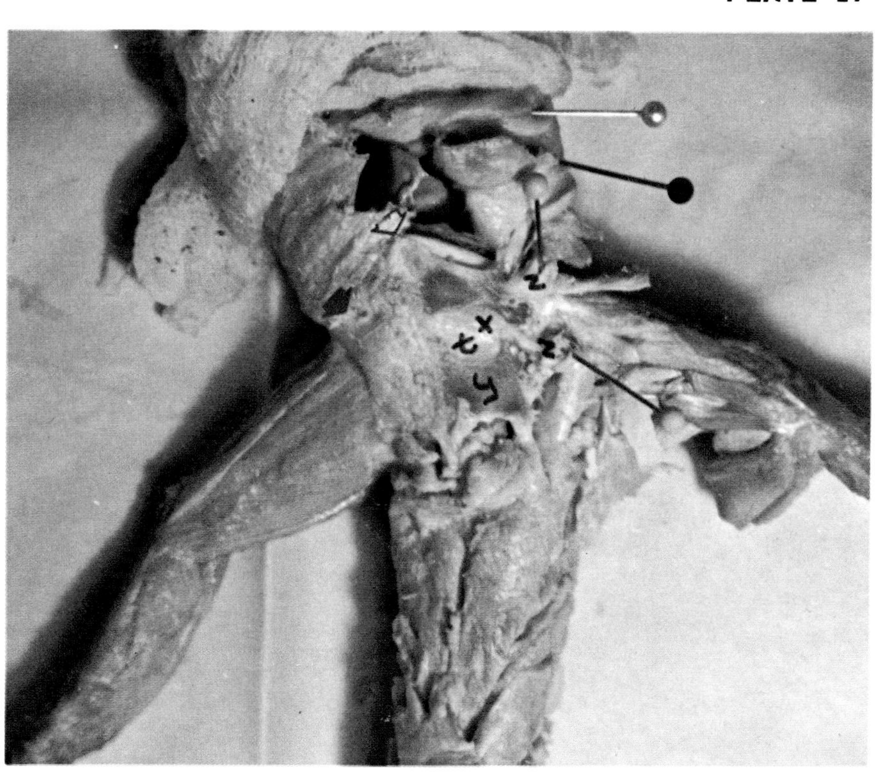

PLATE 18

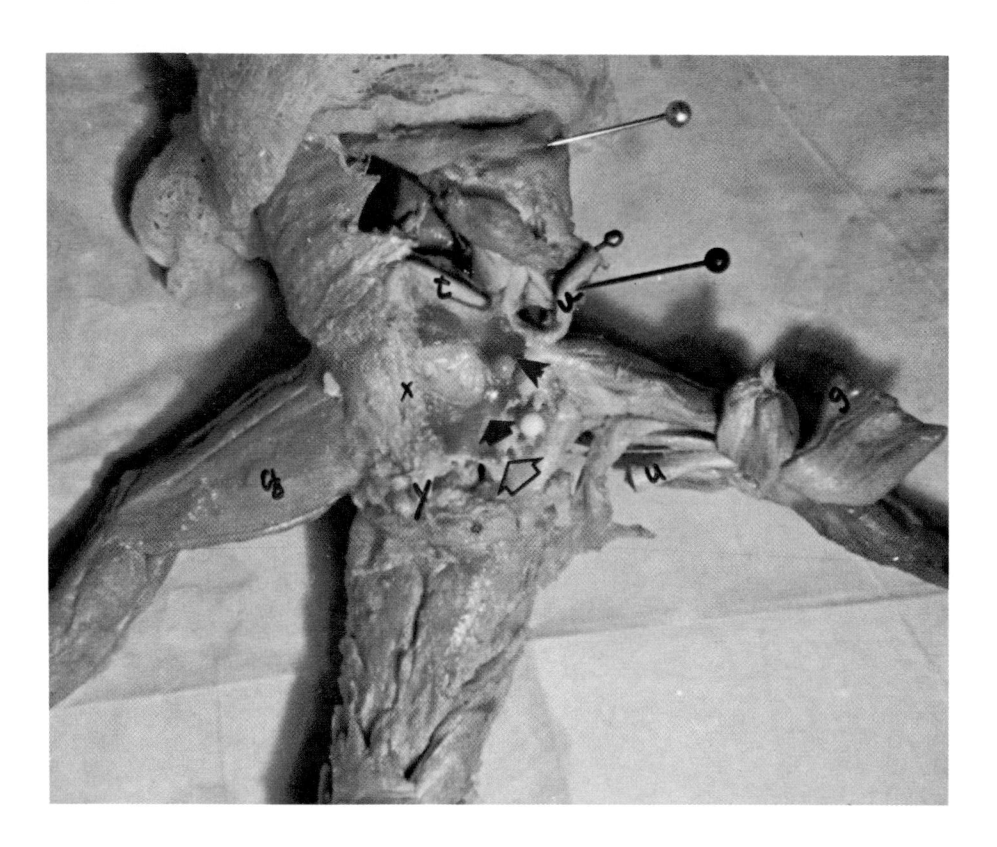

PLATE 18a

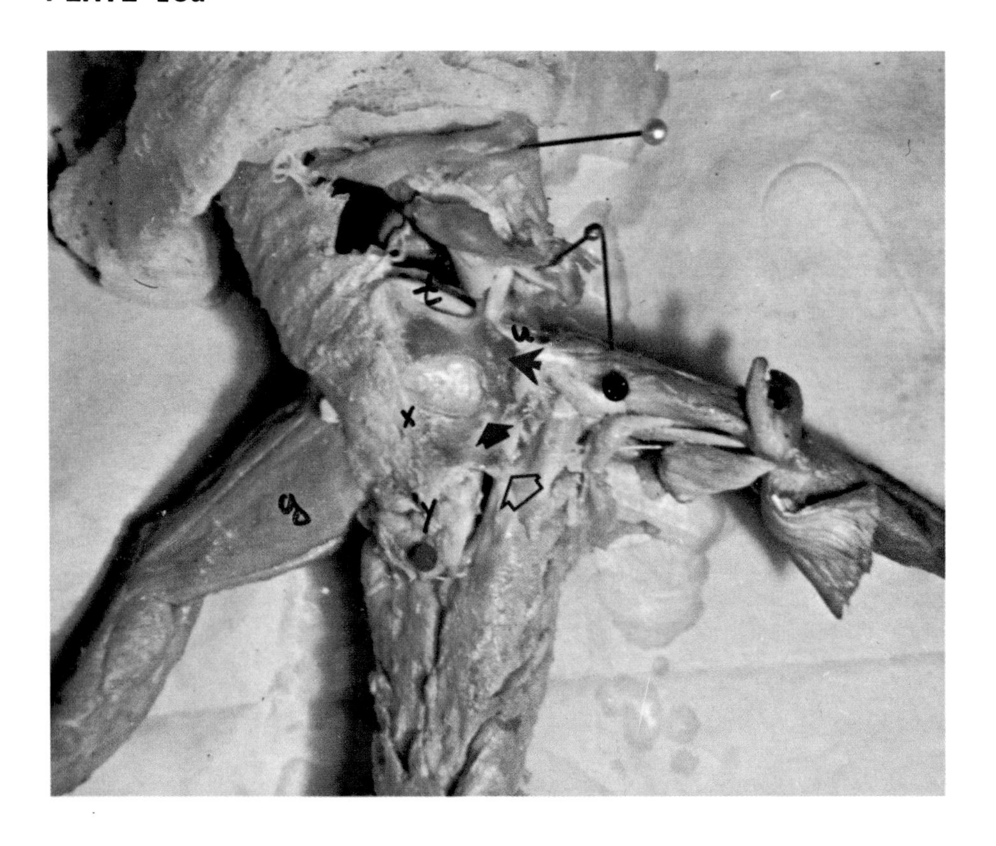

Part of iliopsoas-pectineus removed; pectineus, t.

Area of pubic foramina (thin black arrow head).

Area of ischial foramina (fat black arrow).

Adductor, u, -reflected-blue pin, Plate 18.

Gracilis, g.

Tuberositas ischiadicum (white arrow).

Hypoischiatic process, y.

Pelvic symphysis, x.

Anus-red pin-Plate 18a.

PLATE 19

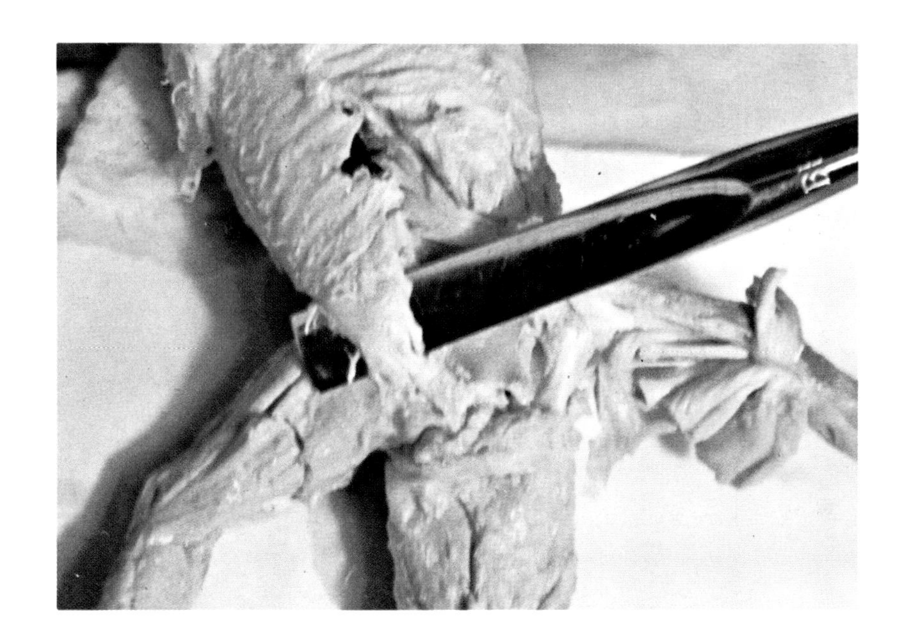

Attachment of obliquus abdominus externus to fascia overlaying ischiopubic plate (forceps beneath).

PLATE 19a

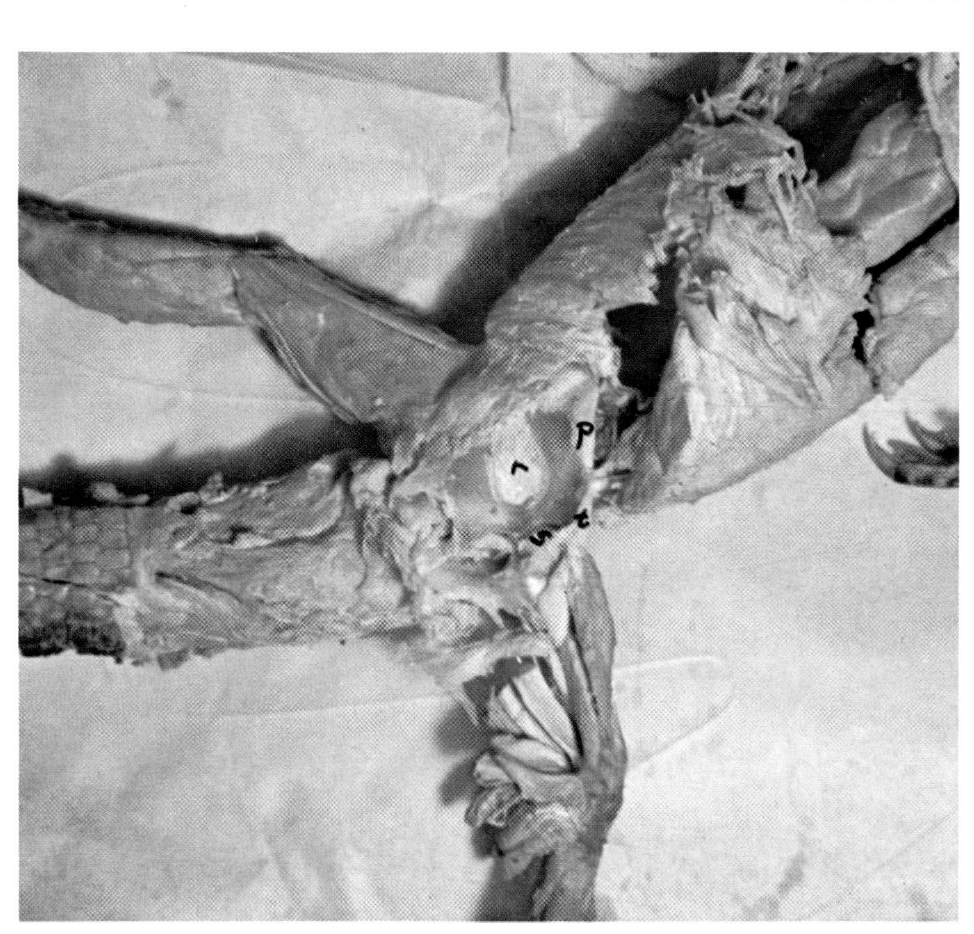

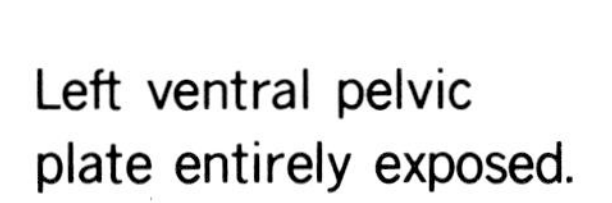

Left ventral pelvic plate entirely exposed.

Area of acetabulum, s.

Pectineus, p.

Iliopsoas-pectineus, t, -from dorsal side of pelvic plate.

Obturator foramen with fascia removed, exposing muscles, r, (of dorsal side of ventral pelvic plate).

PLATE 20

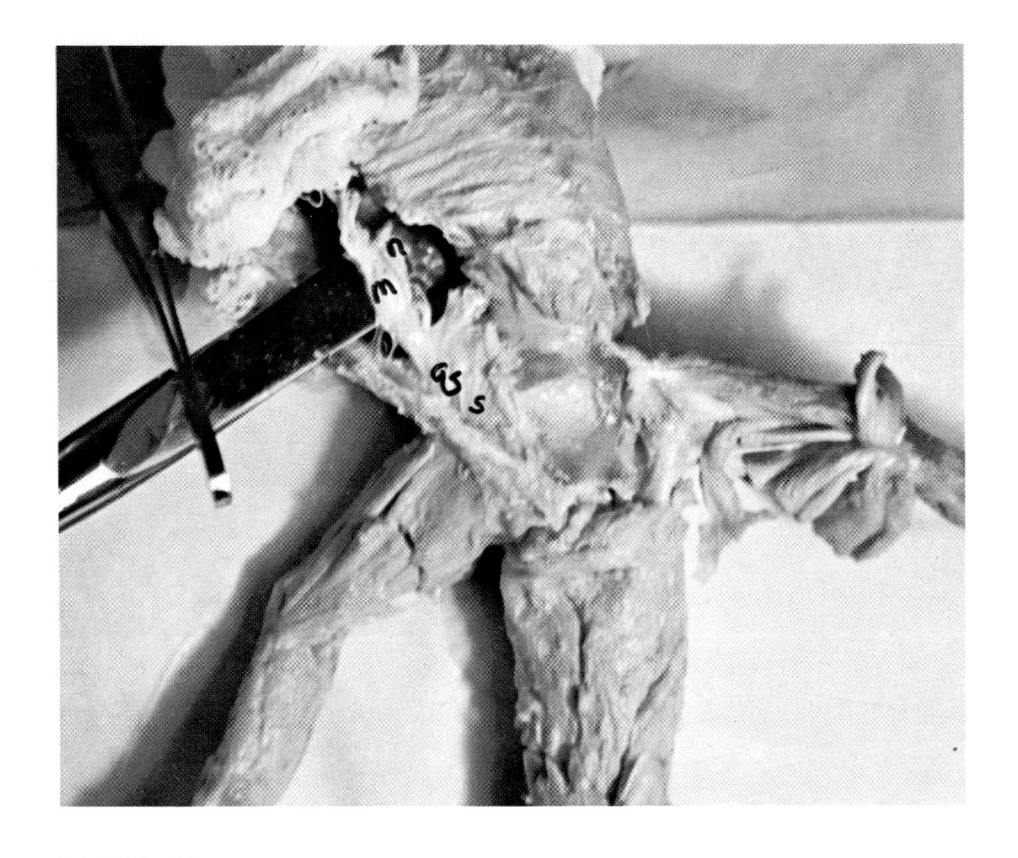

Rectus abdominus (forceps) attached to ischio-pubic plate aponeurosis, q.

Obturator externus, s.

Obliquus abdominus externus, internus, below rectus abdominis, m.

Obliquus abdominus transversus, n.

Obturator internus, v.

PLATE 21

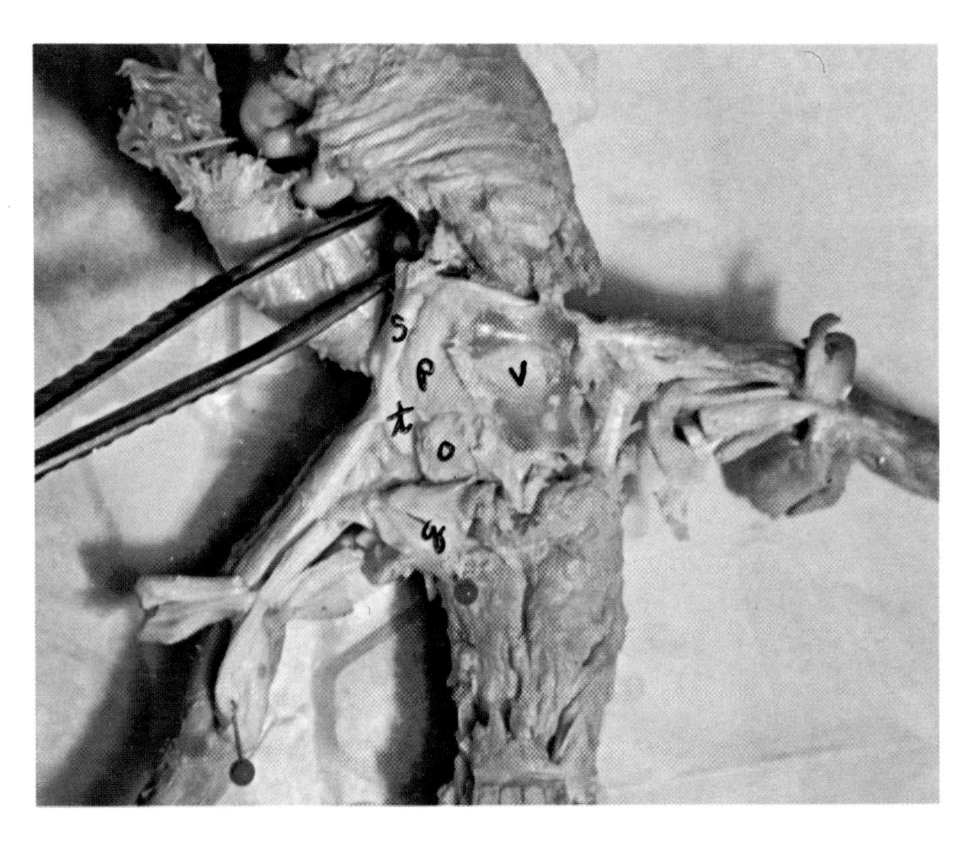

Obturator externus, t, and Quadratus femoris, p.

Gracilis, q, dissected, reflected-red pins.

Piriformis, s.

Gemellus, o.

SPHENODON
PLATES 22-25
Dissection of Dorsal Thigh Muscles
male (actual size)

SPHENODON

Dissection of dorsal thigh muscles in a male rhynchocephalian shown in sequential plates (22-25), demonstrate the stepwise removal of the muscles attached laterally to the os ilium. M. rectus femoris (1) is bisected, (Plate 25), and the "insertion" half is turned back at articularis genu; M. Gluteus (2) is severed at its origin and layed over the M. vastus head (4); M. biceps femoris (3) is seen cut at its belly and the two portions reflected (Plate 24a). This muscle has an origin of only a few fibers on the os ilium lateralis; mostly its fasicles come from its ventral internal basal edge, and the second sacral vertebra (some fibers attached to the postsacral vertebra or caudal). Note that the sheet of the Mm. semimembranosus-semitendinosus complex (9) has fibers originating from the postsacral or caudal vertebrae. Actual size specimen of the male.

Key:
1. Rectus femoris
2. Gluteus
3. Biceps femoris
4. Vastus-which divides at Articularis genu
5. Iliopsoas-pectineus-locus
6. Plexus lumbosacralis (N. peroneus; N. obturatorius)
7. Gastrocnemius
8. Femorococcygeus (Femorocaudalis)
9. Semimembranosus-Semitendinosus complex
10. Obturator internus
11. Sartorius
12. Abdominal cavity (black pigmented peritoneal layer has been broken)
13. Obliquus abdominis externus
14. Os ilium (note backward slant-obtuse angle with the horizontal axis of the body, and the pelvic plate on the ventral side, and the sacral vertebrae)-black outline in ink.

PLATE 22

PLATE 22a

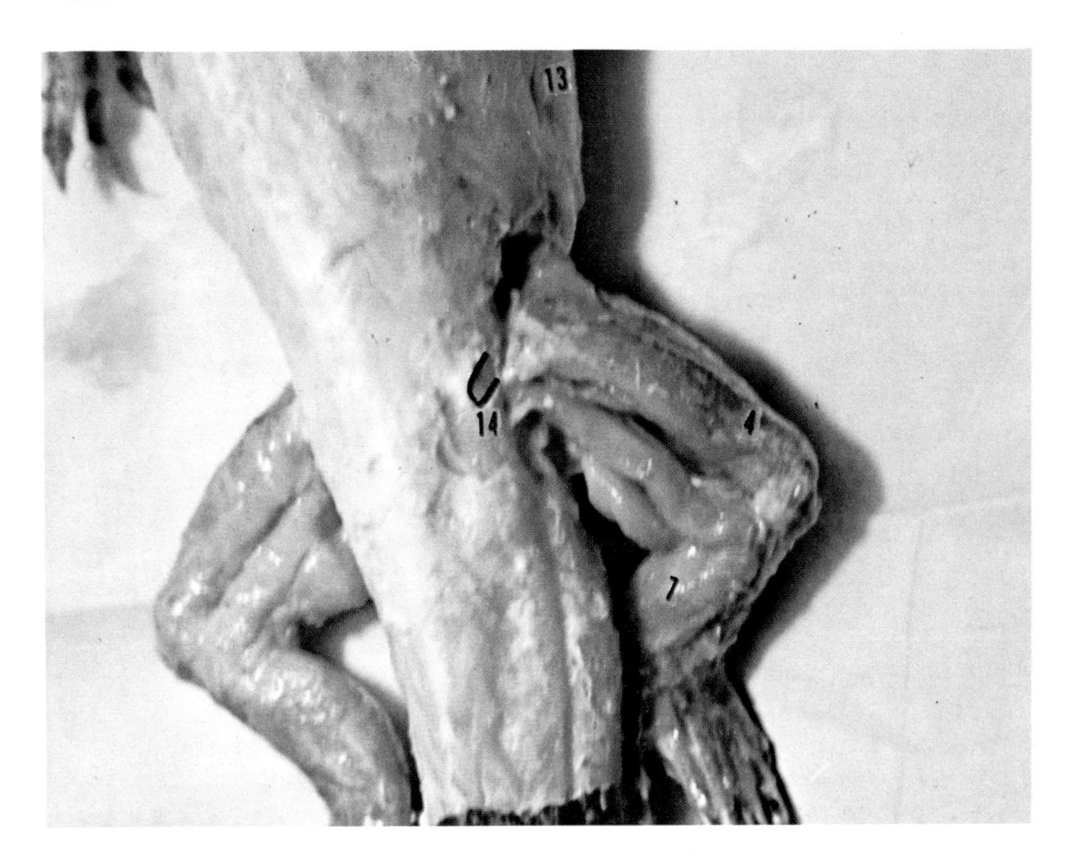

PLATE 23

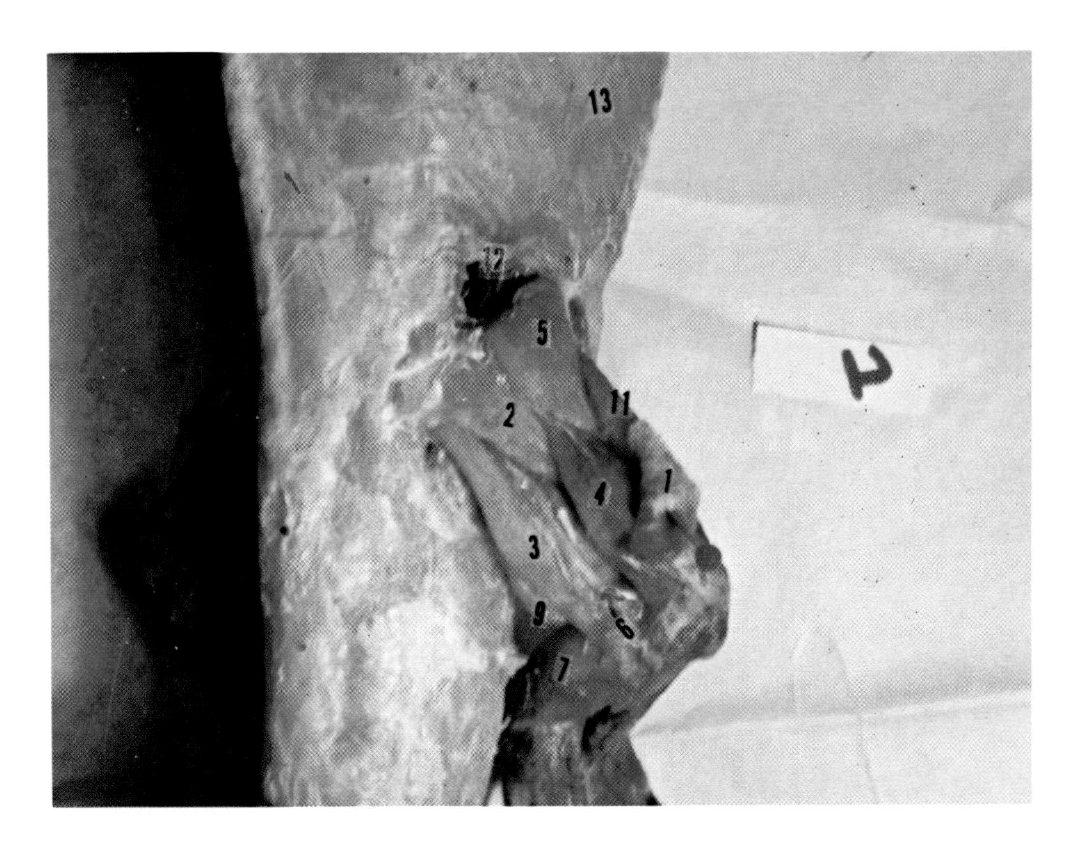

PLATE 23a

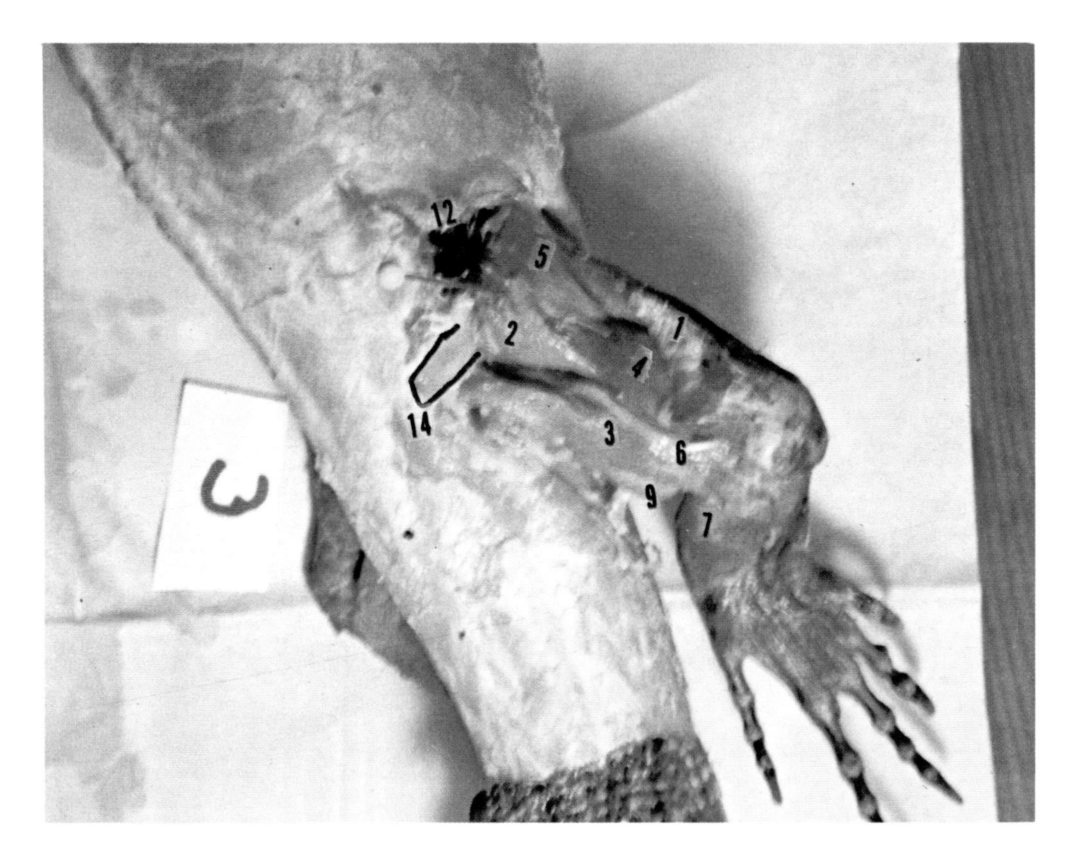

PLATE 24

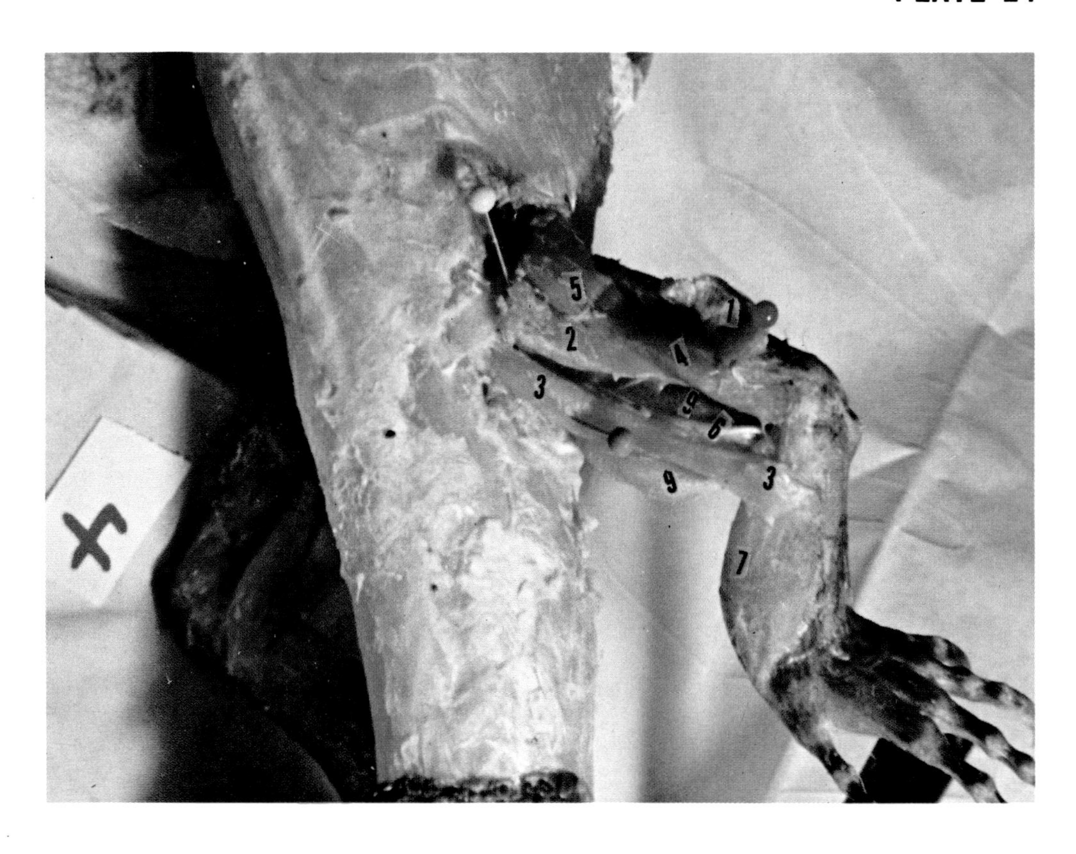

PLATE24a

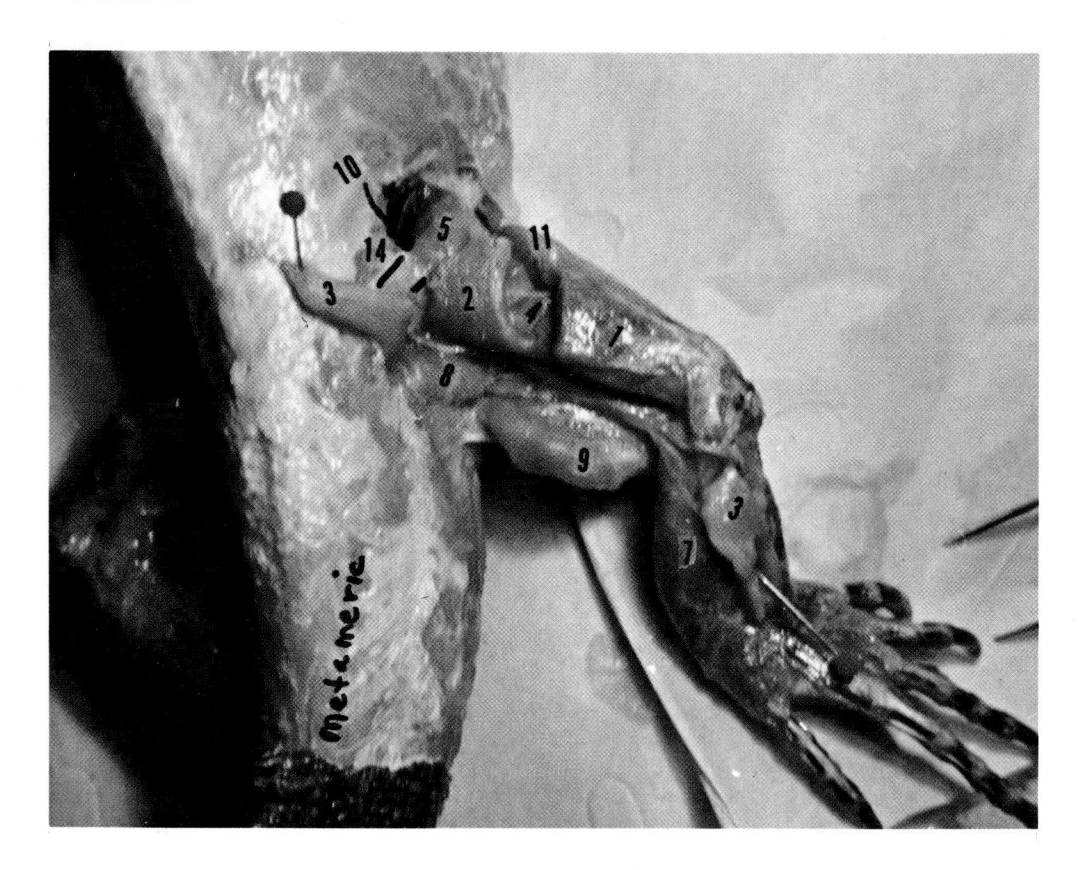

PLATE25

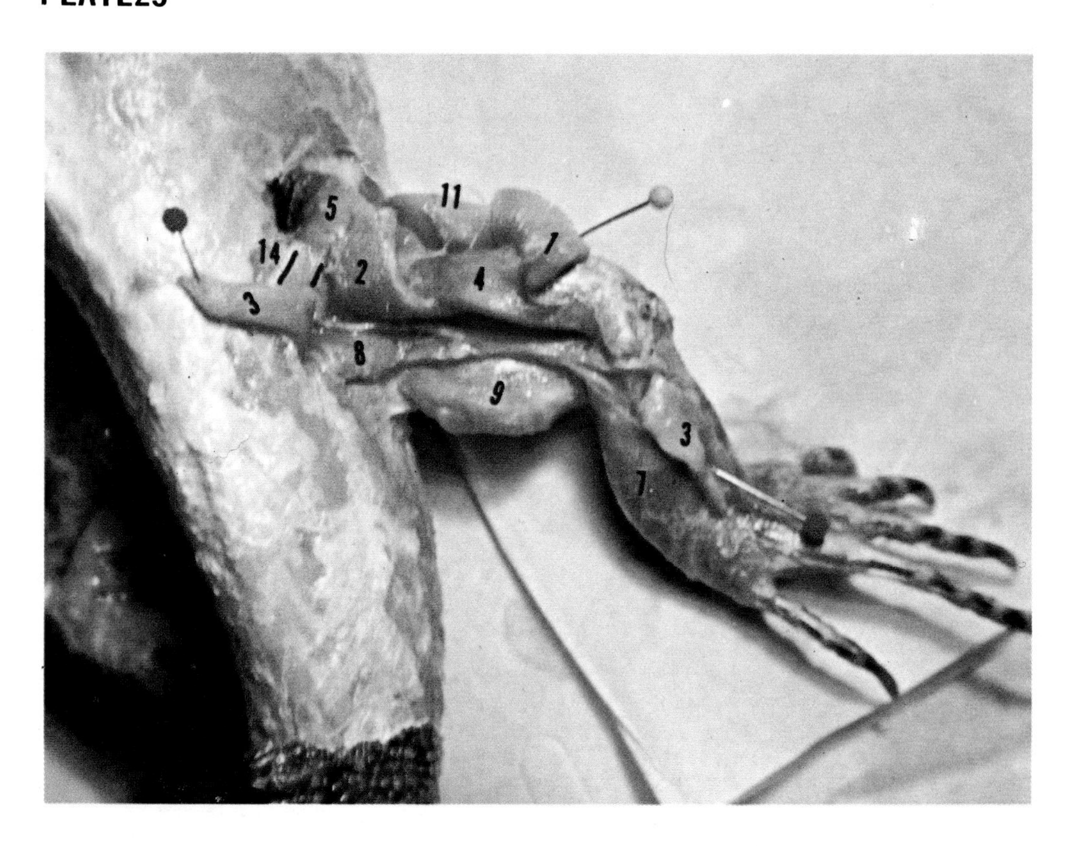

SPHENODON
PLATES 26-29a
Dissection of Dorsal Sacrum and Ilium
female (X's 1½)

Dissection of the Dorsal Sacrum and Ilium:
Their relationship to the Musculus Iliopsoas-pectineus and the dorsal side of the ventral pubic plate of the pelvis.

PLATE 26

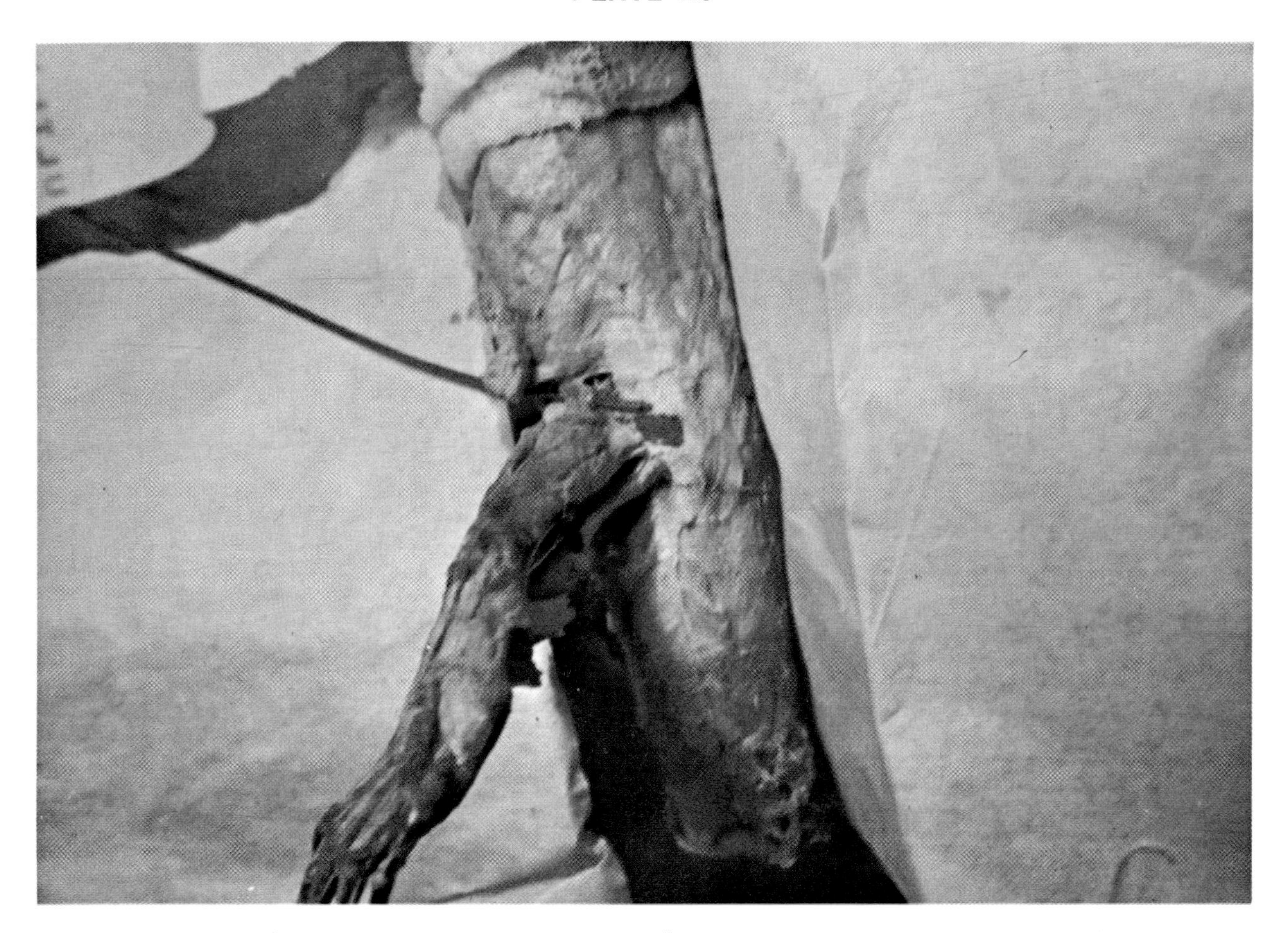

Probe points to the ilium from where muscle origins have been removed.

PLATE 26a

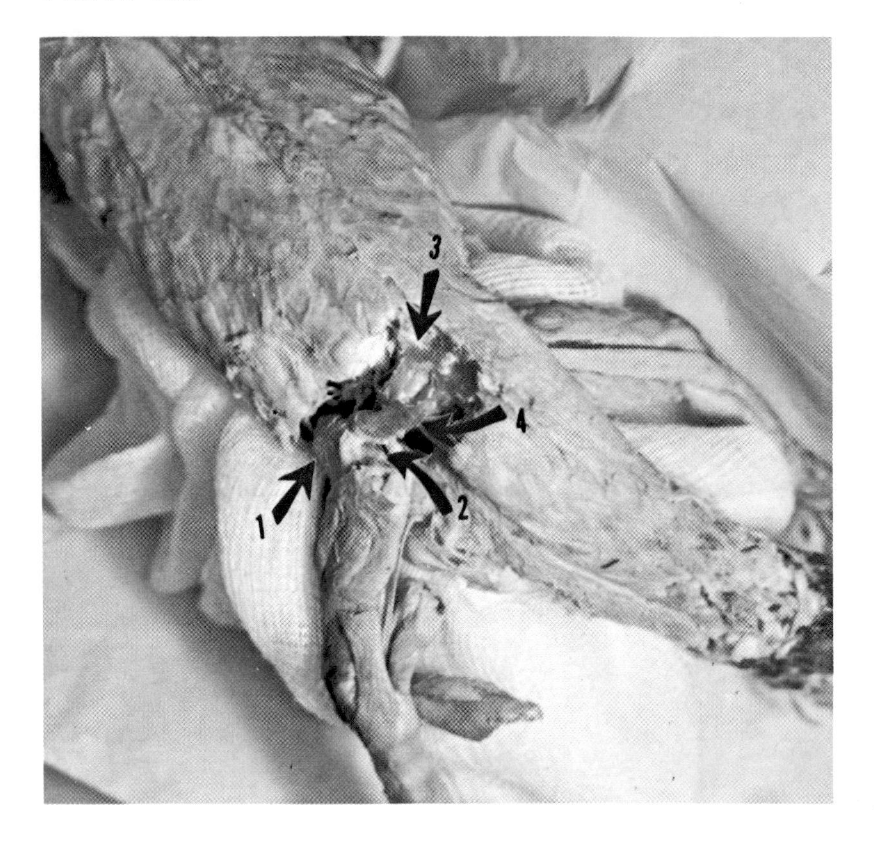

Muscles have been further removed to show the two sacral vertebrae, the ilium, and the M. iliopsoas-pectineus wrapped around the ilium base.

1 - Musculus iliopsoas-pectineus
2 - Head of femur
3 - Sacral vertebrae
4 - Iliac crest

PLATE 27

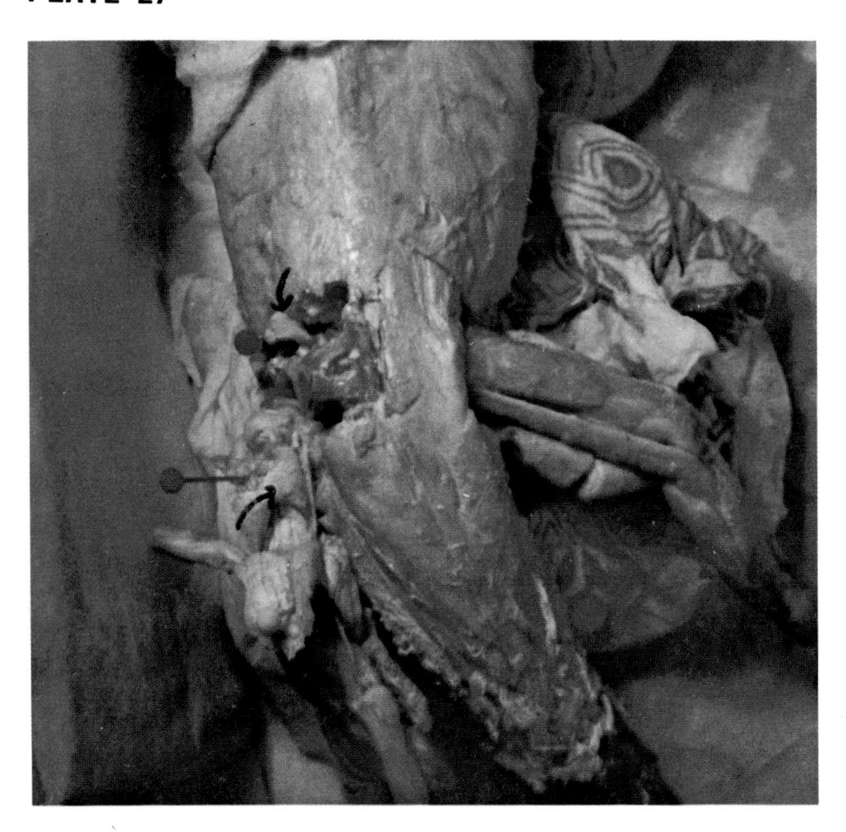

The Musculus iliopsoas-pectineus (arrows) has been bisected and reflected with red pins.

PLATE 28

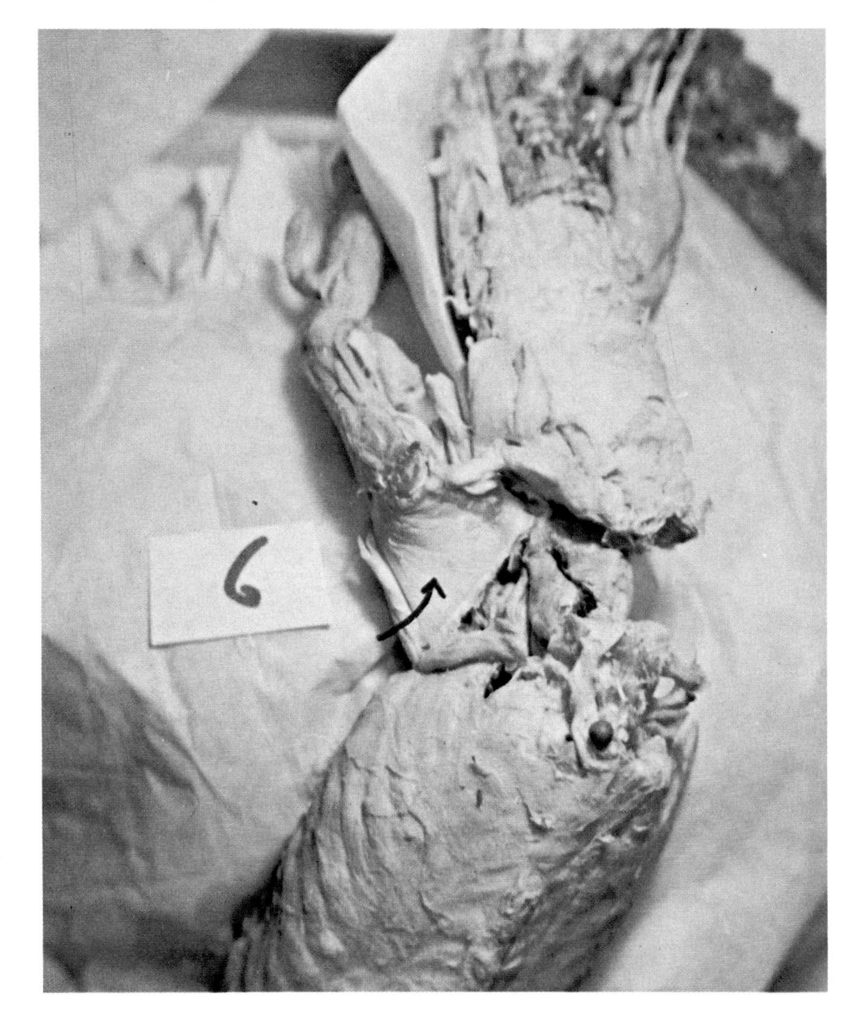

The dorsal side of the ventral pelvic plate showing the muscle bundles (arrow) originating and leading to their insertion, after wrapping around the ilium, onto the lateral side of the proximal femur.

PLATE 28a

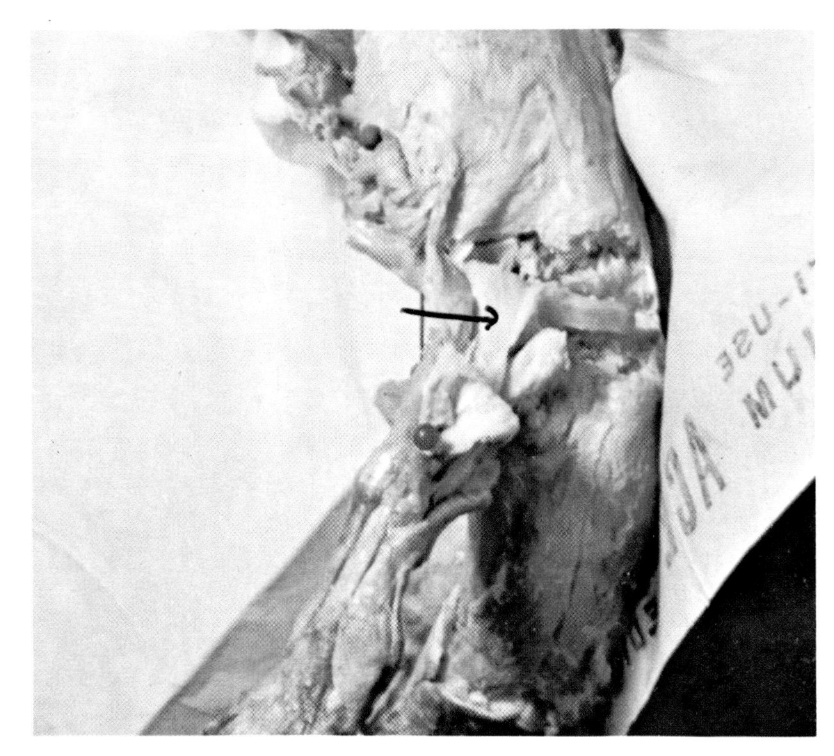

The Musculus iliopsoas-pectineus at ilium base (arrow).

PLATE 29

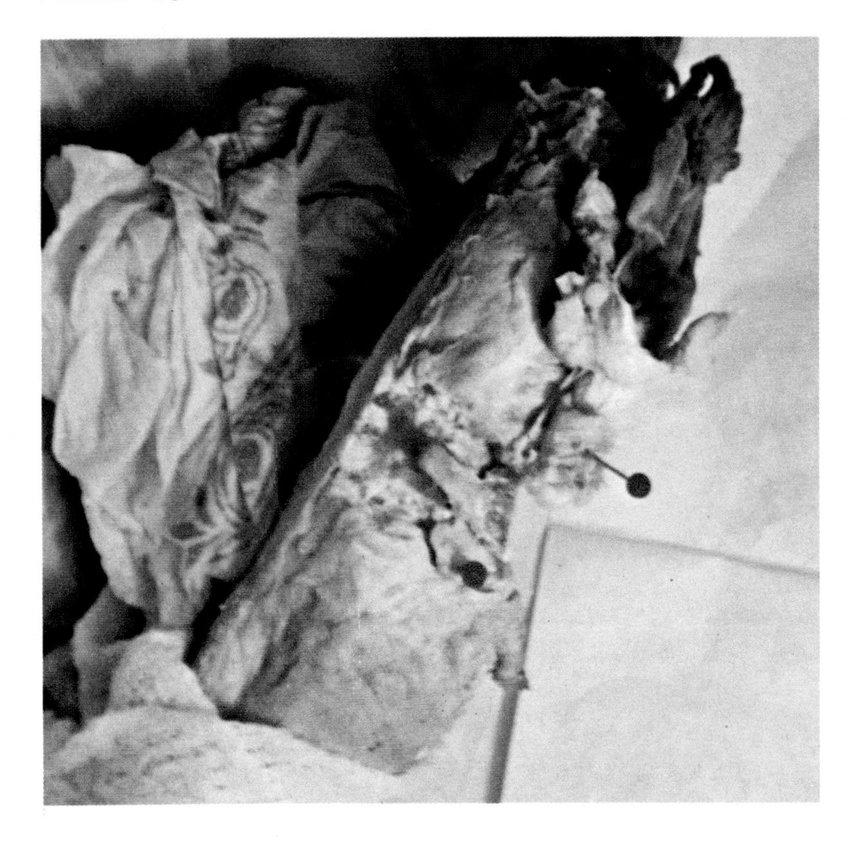

The sacral vertebrae, the ilium, and the left half of the ventral pubic plate of pelvis are shown in relationship to each other. Dorsal view.

PLATE 29a

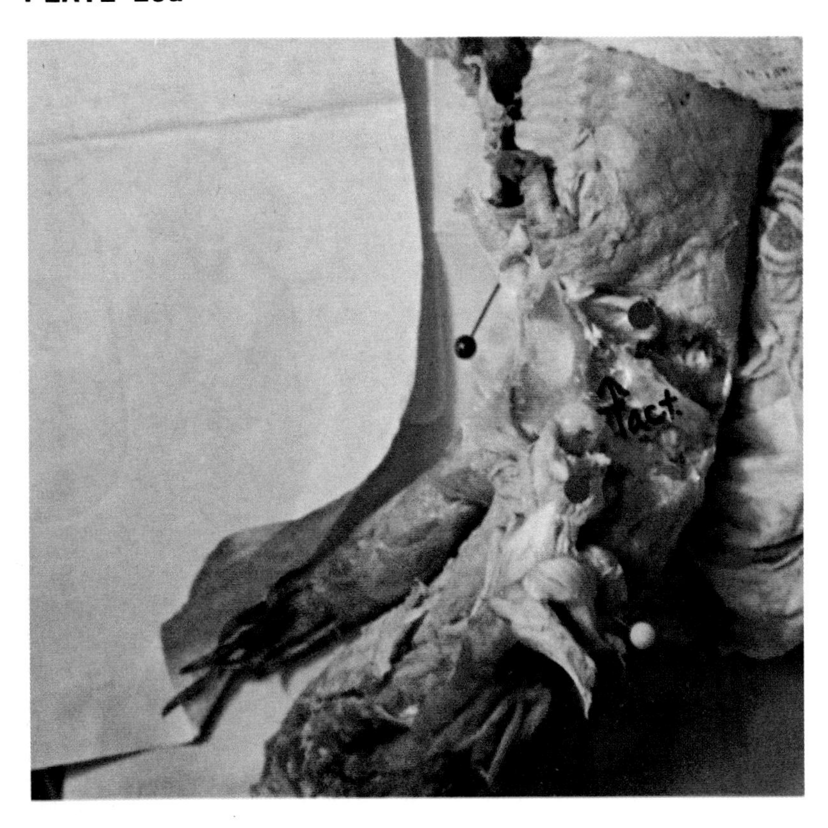

Lateral ventral view of the girdle relationships. Red pins-bisected Musculus iliopsoas-pectineus.